John H. Diedrichs

The Theory of Strains

A Compendium for the Calculation and Construction of Bridges, Roofs and

Cranes...

John H. Diedrichs

The Theory of Strains
A Compendium for the Calculation and Construction of Bridges, Roofs and Cranes...

ISBN/EAN: 9783743465015

Manufactured in Europe, USA, Canada, Australia, Japa

Cover: Foto ©Lupo / pixelio.de

Manufactured and distributed by brebook publishing software
(www.brebook.com)

John H. Diedrichs

The Theory of Strains

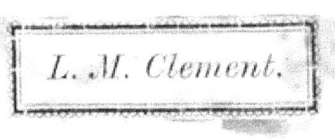
L. M. Clement.

THE

Theory of Strains.

A COMPENDIUM FOR THE

CALCULATION AND CONSTRUCTION OF BRIDGES, ROOFS AND CRANES,

WITH THE APPLICATION OF

TRIGONOMETRICAL NOTES.

CONTAINING

THE MOST COMPREHENSIVE INFORMATION IN REGARD TO THE RESULTING STRAINS FOR A PERMANENT LOAD, AS ALSO FOR A COMBINED (PERMANENT AND ROLLING) LOAD.

IN TWO SECTIONS.

ADAPTED TO THE REQUIREMENTS OF THE PRESENT TIME.

BY

JOHN H. DIEDRICHS,
CIVIL AND MECHANICAL ENGINEER.

ILLUSTRATED BY NUMEROUS PLATES AND DIAGRAMS.

BALTIMORE:
SCHMIDT & TROWE.
1871.

Entered according to Act of Congress, in the year 1871, by
JOHN H. DIEDRICHS,
In the Office of the Librarian of Congress, at Washington.

W`ESTCOTT` & T`HOMSON`,
Stereotypers, Philada.

INSCRIBED TO

WENDEL BOLLMAN, Esq.,

IN TESTIMONY OF HIS INTEREST IN SCIENCE AND ART.

CONTENTS.

	PAGE
PREFACE	7
EXPLANATION OF CHARACTERS USED IN THE CALCULATIONS	11

SECTION I.

A. INTRODUCTION	13
I. THE LEVER	15
II. SUSPENDED WEIGHTS AND THE RESULTING STRAINS	18
III. TRUSSES WITH SINGLE AND EQUALLY-DISTRIBUTED LOAD	21
Suspension Truss Bridge	23
Suspension Bridge	27
B. ROOF CONSTRUCTION	31
C. SEMI-GIRDERS	43
I. SEMI-GIRDERS LOADED AT THE EXTREMITY	43
Cranes	44
II. SEMI-GIRDERS LOADED AT EACH APEX	45
D. GIRDERS WITH PARALLEL TOP AND BOTTOM FLANGES	48
I. STRAIN IN DIAGONALS AND VERTICALS	48
II. STRAIN IN FLANGES	51
III. TRANSFORMATIONS	53
IV. GENERAL REMARKS	53
Directions for the Calculation of Complex Systems	54
E. COMPARATIVE TABLES OF RESULTING STRAINS FOR A PERMANENT LOAD	56
I. SYSTEM OF RIGHT-ANGLED TRIANGLES	56
II. SYSTEM OF ISOSCELES BRACING	58

SECTION II.

GIRDERS CALCULATED FOR COMBINED (PERMANENT AND ROLLING) LOAD.

		PAGE
A.	GIRDERS WITH PARALLEL TOP AND BOTTOM FLANGES	59
	I. THE RIGHT-ANGLED SYSTEM	59
	II. ISOSCELES BRACING	66
	1. Triangular Truss	66
	Calculation of Strains y and u in Diagonals	67
	2. Isometrical Truss	69
	a. Calculation of Diagonals	70
	b. Calculation of Top and Bottom Chords (Flanges)	72
B.	CAMBER IN TRUSSES, WITH PARALLEL TOP AND BOTTOM CHORDS	73
	TABLES CONTAINING THE LENGTH OF ARCHES FOR DEGREES, MINUTES AND SECONDS, FOR A RADIUS AS UNIT	77
C.	PARABOLIC GIRDER OF 48 FEET, OR 16 METRES, SPAN	80
D.	THE ARCHED TRUSS	86
	Calculation of Strain y in the Diagonals	88
	Calculation of Strain in the Verticals V	90
	Transformations	93
E.	THRUST CONSTRUCTION	94
	I. GIRDER 20 FEET IN LENGTH (WITH A SINGLE WEIGHT AT THE CENTRE).	
	II. GIRDER 72 FEET IN LENGTH (CALCULATED FOR PERMANENT AND ROLLING LOAD).	
	Definition of Strain z in the Horizontal Flanges	97
	Definition of Strain y in the Diagonals	98
	Calculation of the Tensile Strains z in the Lower Flanges	99
	Calculation of the Verticals u	101
	III. CALCULATION OF A TRUSS SUSTAINING A DOME	102
	Calculation of Strains x of the Outside Arch	104
	Calculation of Strains z of the Inside Arch	105
	Calculation of the Diagonals y	106

PREFACE.

THE want of a compact, universal and popular treatise on the construction of Roofs and Bridges—especially one treating of the influence of a variable load—and the unsatisfactory essays of different authors on the subject, induced me to prepare the following work.

Bridge-building has been and always will be an important branch of industry, not only to engineers, but also to the masses for the purposes of travel and trade, and, as Colonel Merrill in his recent essay on Bridge-building remarks, "important to railroad companies on account of the large amount of capital invested in their construction."

Bridge literature has often been used by rival parties for the purpose of advancing their own private interests, their motive being competition. Imposing upon the faith and credulity of those whom they pretend to serve, there is no guarantee that worthless structures will not be erected.

Thoroughly independent of any such motive, my aim is to give, especially to bridge-builders and to engineers and architects, the results of my investigation on the subject of calculating strains, in order that capitalists and the public may be benefited and protected.

These calculations will also enable those who have but a limited knowledge of mathematics to acquire the necessary information. For this reason special attention is paid to the arrangement of the work, the whole being made as plain and simple as possible, in order to meet the wants of the common mechanic as well as the experienced engineer.

Though there are many valuable treatises of this kind, there has as yet been no work published serviceable to the degree desired by the practical builder or mechanic—most of the dissertations being too theoretical and hard to comprehend by one not versed in the higher mathematics; and some are so arranged that a clear understanding of the calculations is very difficult.*

The most valuable work in the language is doubtless Mr. Stoney's "Theory of Strains," though the Method of Moments is not developed to that degree which I think necessary for the practical man.

We owe to the renowned German engineers Ritter and Von Kaven the universal application of this *Method* in the work entitled "Dach und Brücken Constructionen," in which it is fully explained by examples and illustrated by diagrams, these being often carelessly neglected in other works.

The above-mentioned work served me very much in the arrangement of this, which I hope will be kindly received.

The work being expressly prepared, as aforesaid, for the use of beginners in the study of mathematics, as well as for the

---

* As an exception, may be named Mr. Shreve's brief but popular treatise in Van Nostrand's "Engineering Magazine," No. xx., August, 1870; Vol. III.

more advanced practical engineer, it will enable them, after a short perusal, to acquire all the necessary information, for which even the trigonometrical notes accompanying the general results are not really required.

The higher classes of colleges and other institutions of learning will find the work very valuable.

On account of the expense, an intended Appendix, containing a rational and concise investigation on "The Strength of Materials," had to be dispensed with; yet I hope with this volume to gratify not only the desire of friends, but to be able with great satisfaction to assist engineers in the pursuit of their high and noble calling.

THE AUTHOR.

OCTOBER, 1870.

EXPLANATION OF CHARACTERS USED IN THE CALCULATIONS.

$=$ Equal, or the sign of equality.
$+$ Plus, or the sign of addition; also, the symbol of positive (tensile) strain.
$-$ Minus, or the sign of subtraction; also, the symbol of negative (compressive) strain.
$\times$ or . Sign of multiplication.
: or $\div$ Sign of division.
, Sign of decimals; also, of thousands.
∞ Sign of infinite.
$<$ Sign of angle, signified by the Grecian cyphers $\alpha, \beta, \gamma, \delta, \epsilon$.
^{2} Sign of square of a number.
$\sqrt{\ }$ Sign of square root of a number.
$^\circ$ Sign of degrees.
$'$ Sign of minutes; also, of feet.
$''$ Sign of seconds; also, of inches.
( ) or [ ] Brackets, to enclose the mathematical expression bound to the same operation.
π The number 3, 14, or periphery for a unit of the diameter.
R Right angle, or 90°.
$\perp$ Vertical.

THE THEORY OF STRAINS.

SECTION I.

A. INTRODUCTION.

To enable the student to comprehend the work, and have a thorough knowledge of certain conditions and examples without studying the whole, it is necessary for him to understand the arrangement of the following pages.

On the first few pages and the appertaining figures at the close of the chapter is found a short description of the lever in its different appliances, the application being only a key to the calculations of strains which follow.

The trigonometrical notes are in many cases almost superfluous. Still, it may be advantageous in this way to accustom the reader to their use. The "Suspended Weights and Resulting Strains" are developed by the parallelogram of forces, and for a plain illustration the results are appended to the figures, which will also be observed on figures of "Trusses with Single and Distributed Load."

In the "Suspended Weights and Resulting Strains" a more elaborate calculation was thought necessary, and therefore an Introduction to the calculation by the "Method of Moments" may be found in its proper place.

This Introduction presents the beginner with a clear and comprehensive knowledge of the formation of Moments; and the equations for Figs. 20, 21, etc., explain the equilibrium of force and leverage.

At the close of this chapter is found the explanation of maximum compressive and tensile strain in the top and bottom chords of a parallel-flanged truss or girder.

On "Roof Construction" (B), Plates 6 to 11, remarks are unnecessary. The builder can with ease find from the figures a system to suit his purpose. (See also "Arched Trusses," D, Section II.)

On "Semi-Girders" (C) the calculation of strains is treated in the way heretofore generally known (determining from the centre toward the abutment), after which the "Method of Moments" is applied to the same example, followed by a more elaborate explanation of the principles of Moments on the crane skeleton (Plate 14), whose single members are altogether divergent.

The thorough calculation of a truss with horizontal top and bottom flanges (right-angled system D, Sect. I.), with the resulting strains for a system of braces, all of which are inclined in the same direction, shows how easily by transformation of the strains a system of bracing just reversed can be formed. (Comp. D, Sect. II.)

From the comparative tables, E, I., II. (Plate 18, 19) those who are not mathematicians can find, by a little study (for an assumed load "W"—a variable load not considered), the strains in flanges, braces and ties. (See D, IV., Sect. I.)

The progress of panels, and by this the increase of stress in the different members, are *ad libitum* to be extended (are optional).

Where no composite strains appear, in the skeletons double lines make the compressive (—) strains more obvious, the tensile (+) strains being always represented by single lines. The assumed load in the calculations is equally divided on the apexes; but in general some attention may be paid to a peculiar load—say from a single locomotive—at a certain apex, this being observed in examples on "Suspension Truss" (III., Sect. I.).

The calculations in Sect. II. with regard to the influence of a variable load are more difficult to understand. Still, by the results of strains in the skeletons it is easy enough to form an idea about this matter, and to see the importance of counter-bracing or tying at centre of railroad-bridge trusses.

What experience and observation have already taught to the practical railroad man is here *fully* shown by figures.

Information is given on parallel-flanged trusses for the so-called "Camber in Bridges" at B. Sect. II., which to many builders has heretofore been only a matter of experiment.

Yet it is to be remarked that for the calculations E, Sect. II., Plates

29 to 34, a variation in
will be perceived, the
being the horizontal
vertex (centre), an
ments.

Farther expl
referred to.

The calcul
nections at
relieving th
strain. T
ture are

Pla
Fi
g. 4.

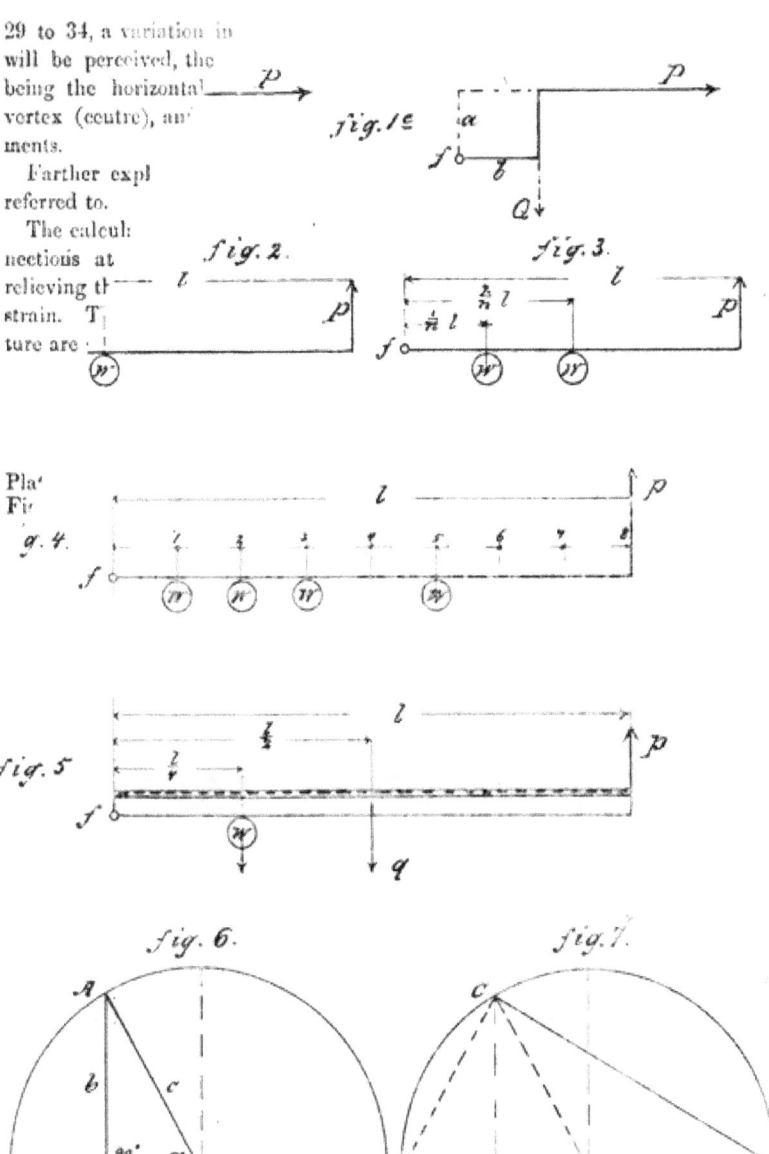

necessary. The builder can with ease ... system to suit his purpose. (See also "A from the fulcrum equal tion II.)

On "Semi-Girders" (C) the calculation of ... the way heretofore generally known (determin... toward the abutment), after which the "Method applied to the same example, followed by a more c... nation of the principles of Moments on the crane s... 14), whose single members are altogether divergent. ... is divided in

The thorough calculation of a truss with horizont... bottom flanges (right-angled system D, Sect. I.), with th... strains for a system of braces, all of which are inclin... same direction, shows how easily by transformation of th... a system of bracing just reversed can be formed. (Co... Sect. II.)

From the comparative tables, E, I., II. (Plate 18, 19) those ... are not mathematicians can find, by a little study (for an assu... load "W"—a variable load not considered), the strains in flanged, braces and ties. (See D, IV., Sect. I.)

The progress of panels, and by this the increase of stress in the different members, are ad libitum to be extended (are optional).

Where no composite strains appear, in the skeletons double lines make the compressive (—) strains more obvious, the tensile (+) strains being always represented by single lines. The assumed load in the calculations is equally divided on the apexes; but in general some attention may be paid to a peculiar load—say from a single locomotive—at a certain apex, this being observed in examples on "Suspension Truss" (III., Sect. I.).

The calculations in Sect. II. with regard to the influence of a variable load are more difficult to understand. Still, by the results of strains in the skeletons it is easy enough to form an idea about this matter, and to see the importance of counter-bracing or tying at centre of railroad-bridge trusses.

What experience and observation have already taught to the practical railroad man is here *fully* shown by figures.

Information is given on parallel-flanged trusses for the so-called "Camber in Bridges" at B, Sect. II., which to many builders has heretofore been only a matter of experiment.

Yet it is to be remarked that for the calculations E, Sect. II., Plates

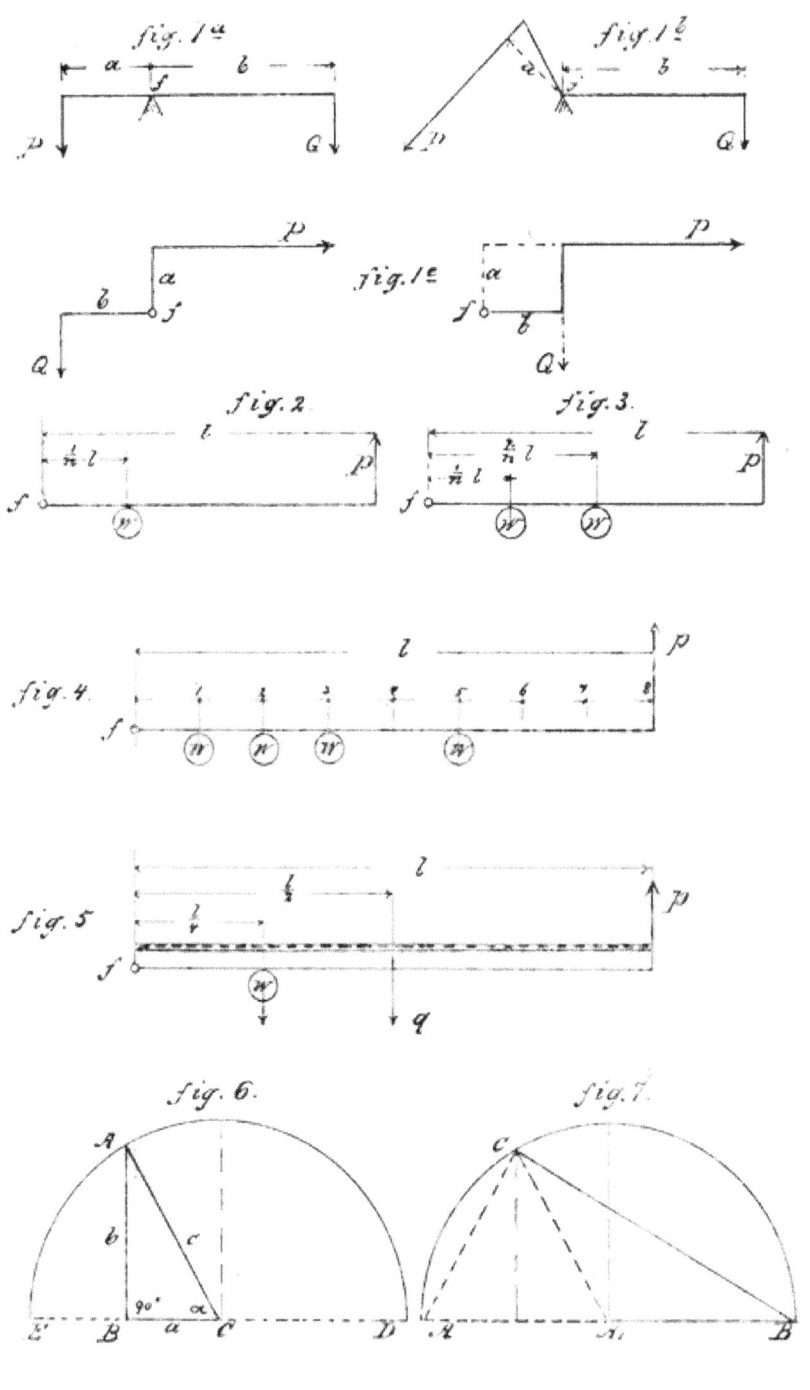

In every triangle the sum of enclosed angles,

$$2R. = 2 \times 90°,$$

6.] so in the right angled triangle ABC, Fig. 6, the angles A and C together $= 90° = R$, because angle $B = 90°$.

In Trigonometry we say—

$\frac{b}{c} =$ sine α; $\frac{b}{a} =$ tangent α; $\frac{c}{a} =$ secant α.

$\frac{a}{c} =$ cosine α; $\frac{a}{b} =$ cotangent α; $\frac{c}{b} =$ cosecant α; or, contracted,

sin., cos., tang. or tg., cotang. or cotg., sec. and cosec.

For a radius, AC, as a unit, the line AB simply is called sine; the central distance, BC, cosine; and BE the versed sine of the angle α.

For certain angles α the relation $\frac{b}{c}, \frac{b}{a}, \frac{c}{a}$, etc., have certain and distinct numerical values. (See Haslett's or Haswell's "Tables of natural sines, cosines, tangents and cotangents from 1 to 90 degrees.")

Each triangle, ABC, A_1BC (Fig. 7), consists of six members—i.e., three sides and three angles, from which always three are dependent on the rest; therefore, when three out of these six members are known, we can construct, or with more exactness we can calculate, the others, provided one at least of the given parts is a side.

For the transformation of trigonometrical functions, short notices in the form of a table, also the numerical values (natural sin, cos, etc.) of the principal angles, may be serviceable, viz.:

$$\sin x = \frac{1}{\operatorname{cosec} x} = \frac{\tang x}{\sec x} = \sqrt{1 - \cos^2 x}.$$

$$\cos x = \frac{1}{\sec x} = \frac{\cotang x}{\operatorname{cosec} x} = \sqrt{1 - \sin^2 x}.$$

$$\tang x = \frac{1}{\cotang x} = \frac{\sin x}{\cos x} = \sqrt{\sec^2 - 1}.$$

$$\cotang x = \frac{1}{\tang x} = \frac{\cos x}{\sin x} = \sqrt{\operatorname{cosec}^2 x - 1}.$$

$$\sec x = \frac{1}{\cos x} = \frac{\operatorname{cosec} x}{\cotang x} = \sqrt{1 + \tang^2 x}.$$

$$\operatorname{cosec} x = \frac{1}{\sin x} = \frac{\sec x}{\tang x} = \sqrt{1 + \cotang^2 x}.$$

Degrees.	Sin.	Cos.	Tang.	Cotang.
0	0	1	0	∞
15	0,258	0,965	0,267	3,732
20	0,342	0,939	0,363	2,747
25	0,422	0,906	0,466	2,144
30	$\tfrac{1}{2} = 0{,}5$	$\tfrac{1}{2}\sqrt{3} = 0{,}866$	$\tfrac{1}{3}\sqrt{3} = 0{,}577$	$\sqrt{3} = 1{,}732$
40	0,642	0,766	0,839	1,191
45	$\tfrac{1}{2}\sqrt{2} = 0{,}707$	$\tfrac{1}{2}\sqrt{2} = 0{,}707$	1,00	1,00
50	0,766	0,642	1,191	0,839
60	$\tfrac{1}{2}\sqrt{3} = 0{,}866$	$\tfrac{1}{2} = 0{,}5$	$\sqrt{3} = 1{,}732$	$\tfrac{1}{3}\sqrt{3} = 0{,}577$
65	0,906	0,422	2,144	0,466
75	0,965	0,258	3,732	0,267
$R = 90$	$+1$	0	$+\infty$	0
$2R = 180$	0	-1	0	$-\infty$
$3R = 270$	-1	0	$-\infty$	0
$4R = 360$	0	$+1$	0	$+\infty$

$\sin(R - x) = {}^+\cos x$, and $\cos(R - x) = {}^+\sin x$.
$\sin(R + x) = {}^+\cos x$, and $\cos(R + x) = {}^-\sin x$.

II. SUSPENDED WEIGHTS AND THE RESULTING STRAINS.

Plate 2,] In Fig. 8^a, when $W = 5000$ lbs, $ab = be = 10$,
Fig. 8^a, $bc = 8$, and $ac = ec = 12{,}8$ feet; the vertical strain at
" 8^b.] c on each string $= \dfrac{W}{2} = \dfrac{5000}{2}$.

And, further, the actual strain in the direction of the string,

$$p = q = \frac{W}{2} \cdot \frac{ac}{bc}.$$

or $\qquad p = q = \dfrac{5000}{2} \times \dfrac{12{,}8}{8} = 4000$ lbs.

All other information is given in Fig. 8^b.

When a heavy body, $ABCD$ (Fig. 8^c), is suspended by two
8^c.] oblique strings, DH and CH, in a vertical plane, a straight
line drawn through the intersection will pass through the
centre of gravity, G, of the body.

9.] For the force in the direction ad, represented by q, we find from Fig. 9,

$$gh = ab \cdot \frac{bd}{L} = ab \cdot \frac{W}{L};$$

$$W = gh + id = gh + q \cdot \cos \alpha;$$

$$W = ab \cdot \frac{W}{L} + q \cdot \frac{bd}{ad};$$

or

$$q = W \cdot \left(1 - \frac{ab}{L}\right) \cdot \frac{ad}{bd} = W \cdot \frac{L - ab}{L} \cdot \frac{ad}{bd};$$

i.e.,

$$q = W \cdot \frac{bc}{L} \cdot \frac{ad}{bd};$$

and in the same manner from similarity of triangles,

$$p = W \cdot \frac{ab}{L} \cdot \frac{cd}{bd}.$$

In the equation for q is $W \cdot \frac{bc}{L}$, the vertical strain at d for the string ad; in the second equation is $W \cdot \frac{ab}{L}$, the vertical strain at d for the string cd,—equal to the shearing strains V and V_1 on the supports.

Example.—When, again,

$W = 5000$ lbs., $L = 100$ feet,
$ab = 10'$, $bc = 90'$, and $bd = 8'$,
$ad = 12.84'$, and $cd = 90.35'$,

so

$$p = 5000 \times \frac{10}{100} \times \frac{90.35}{8} = 5646 \text{ lbs.},$$

and

$$q = 5000 \times \frac{90}{100} \times \frac{12.84}{8} = 7029 \text{ lbs.};$$

then the results for the horizontal strain x and the vertical strain V at the right support are—

$$x = p \cdot \frac{cb}{cd} = p \cdot \sin \beta, \text{ or } x = W \cdot \frac{ab}{L} \cdot \frac{bc}{bd},$$

and

$$V = p \cdot \frac{bd}{cd} = p \cdot \cos \beta, \text{ or } V = W \cdot \frac{ab}{L};$$

thus

$$x = 5000 \times \frac{10}{100} \times \frac{90}{8} = 5625 \text{ lbs.};$$

$$V = 5000 \times \frac{10}{100} = 500 \text{ lbs.}$$

The results for the horizontal strain x_1 and the vertical strain V_1 at the left support are—

$$x_1 = W \cdot \frac{bc}{L} \cdot \frac{ab}{bd},$$

and
$$V_1 = W \cdot \frac{bc}{L};$$

thus
$$x_1 = 5000 \times \frac{90}{100} \times \frac{10}{8} = 5625 \text{ lbs.},$$

and
$$V_1 = 5000 \cdot \frac{90}{100} = 4500 \text{ lbs.};$$

therefore, also, $x = x_1$,

for
$$W \cdot \frac{ab}{L} \cdot \frac{bc}{bd} = W \cdot \frac{bc}{L} \cdot \frac{ab}{bd}. \qquad (\text{Fig. 17.})$$

10.] For Fig. 10, when $W = 5000$ lbs.; ab, ad and bd the same as before—

$$Y = W \cdot \frac{ad}{bd} = 5000 \times \frac{12.84}{8} = 8025 \text{ lbs.},$$

and
$$Y_1 = W \cdot \frac{ab}{bd} = 5000 \times \frac{10}{8} = 6250 \text{ lbs.}$$

11.] In Fig. 12,
$$P : Q : R = \sin . \, cdb : \sin . \, adb : \sin . \, ade.$$

12.] In general, for every triangle,
$$y = x + z, \qquad (\text{Fig. 12.})$$
and, as here, $\quad x + y = z + m = R,$
$$x + x + z = z + m,$$
or, $\quad m = 2x.$

Also, from similarity of triangles ABC, ADC and DBC,
$$\frac{c}{b} = \frac{b}{a}, \quad \text{or } c = \frac{b^2}{a};$$

and as $\quad AB = c + a, \quad$ the diameter $= \frac{b^2}{a} + a$;

i. e., the diameter equals the square of one-half the chord divided by the height of the arc added to the height of the arc.

The height of the arc CB results from the chord CB in the same way.

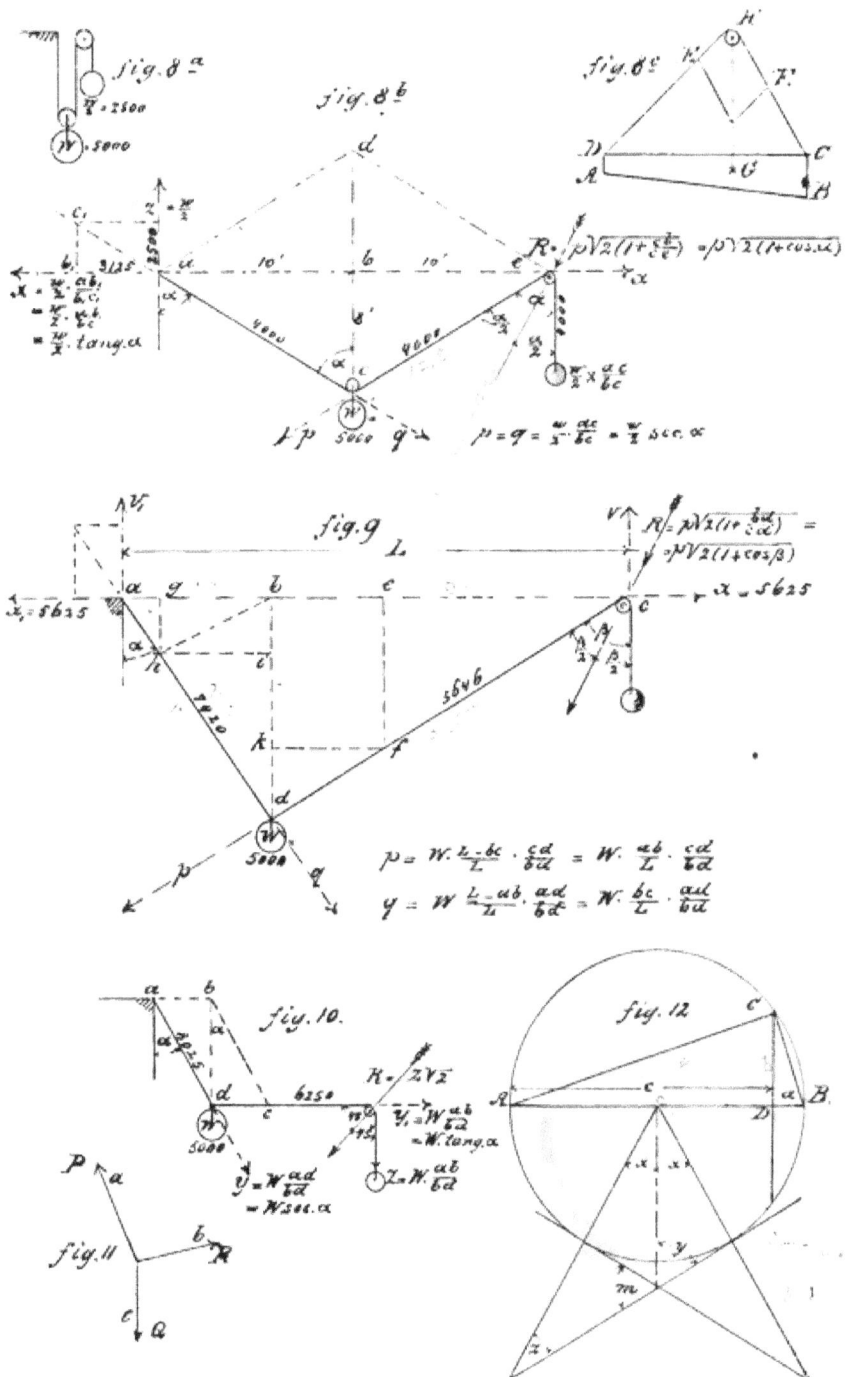

Application in "Camber in Bridges," B, Sect. II. There, also, the geometrical rule,

$$\frac{\text{arc}}{\text{circumference}} = \frac{\text{angle at centre}}{360°}$$

[Plate 2—Figs. 8 to 12.]

III. *TRUSSES WITH SINGLE AND EQUALLY-DISTRIBUTED LOAD.*

Plate 3,]   A most frequent structure is the trussed beam (Fig.
Fig. 13.]   13).

The post at the centre is called the king-post.

14.]   The whole is a combined system, in which the horizontal beam, according to its stiffness, relieves the tie-rods from an aliquot amount of strain.

For the greatest exertion to which the tie-rods in the most unfavorable case could be exposed, we may use the result from Fig. 8^b, under the supposition that the horizontal beam counteracts only the horizontal forces. (For instance, when butted at the centre.)

To compute the stress we have the following:

$$p = q = \frac{W}{2} \cdot \frac{ac}{bc} = \frac{W}{2} \cdot \sec \alpha,$$

for a single weight at centre.

Example.—The assumed weight $= 20000$ lbs.; between supports, 24 feet.

The length $bc = 5,59'$, and $ac = 13,24$ (which can be measured near enough for most purposes from a skeleton);

$$p = q = \frac{W}{2} \cdot \frac{ac}{bc} = 10,000 \times \frac{13,24}{5,59} = 23,700 \text{ (approx.)}.$$

The angle α being in this case $65°$, for a calculation, using the preceding tables,

$$p = q = \frac{W}{2} \cdot \sec \alpha = \frac{20000}{2} \times \frac{1}{\cos \alpha} = 10000 \times \frac{1}{0,422} = 23700,$$

$$bc = ab \cdot \tan 25° = 12 \times 0,4663 = 5,595,$$

$$ac = ab \cdot \sec 25° = ab \cdot \frac{1}{\cos 25°} = 12 \times \frac{1}{0,906} = 13,24.$$

The vertical pressure in the king-post under the supposition before $= 20,000$ lbs.; at each support $= 10,000$ lbs.; and the compression in the horizontal beam,

$$H = \frac{W}{2} \cdot \frac{ab}{bc} = 10000 \times \frac{12}{5,59} = 21467 \text{ lbs.}$$

15ª.] When in Fig. 15ª the strains p or q and the angle γ are known, we find the resulting vertical strain, R, also by means of the parallelogram of forces, viz.,

$$R = \sqrt{p^2 + q^2 + 2 p \cdot q \cdot \cos \gamma};$$

or, because in this case $\alpha = \beta$, therefore, $p = q$;

also, $\gamma = 2.65° = 130°$, or $= R + 40$,

and $\cos (R + 40) = - \sin 40$ (see preceding table),

so $R = \sqrt{2 p^2 + 2 p^2 (-\sin 40)} = \sqrt{2 p^2 (1 - \sin 40)}$,

15ᵇ.] $R = \sqrt{2 \times 23700^2 (1 - 0,642)} = 20000$ lbs.*

For a structure, Fig. 15ᵇ, reserved to the preceding one (15ª), the numerical value of strains is quite the same, but of opposite character, provided the enclosed angles are the same.

16.] If, as in Fig. 16, the load $= 40000$ lbs., equally distributed on the beam, then each support will sustain again one-half of the load; but the reaction, D, of each support will be only one-quarter of the load $= 10000$ lbs.; and for the same exertion a truss or beam, charged with an equally-distributed load, will sustain twice as much as when loaded with a single weight at the centre. (Comp. Fig. 14.)

The distribution of forces on supports and at the centre is explained by Fig. 16.

For an angle, $\alpha = 63° 26'$, or, also, $\beta = 26° 34'$, the depth of the truss being always equal to one-fourth of its length, for $\tang. 26° 34' = 0,5$.

Now, $\frac{bc}{ab} = \tang. \beta$, or, $bc = db \cdot 0,5 = 12 \times 0,5 = 6$,

---

* For the angle $acd = d$ in Fig. 15ᵇ,

$R = \sqrt{p^2 + q^2 + 2 pq \cos d}$, and for the angle $edS = e$,

$R = \sqrt{p_1^2 + p^2 - 2 p_1 \cdot p \cdot \cos e}$. (See Fig. 8ᵇ and Fig. 9.)

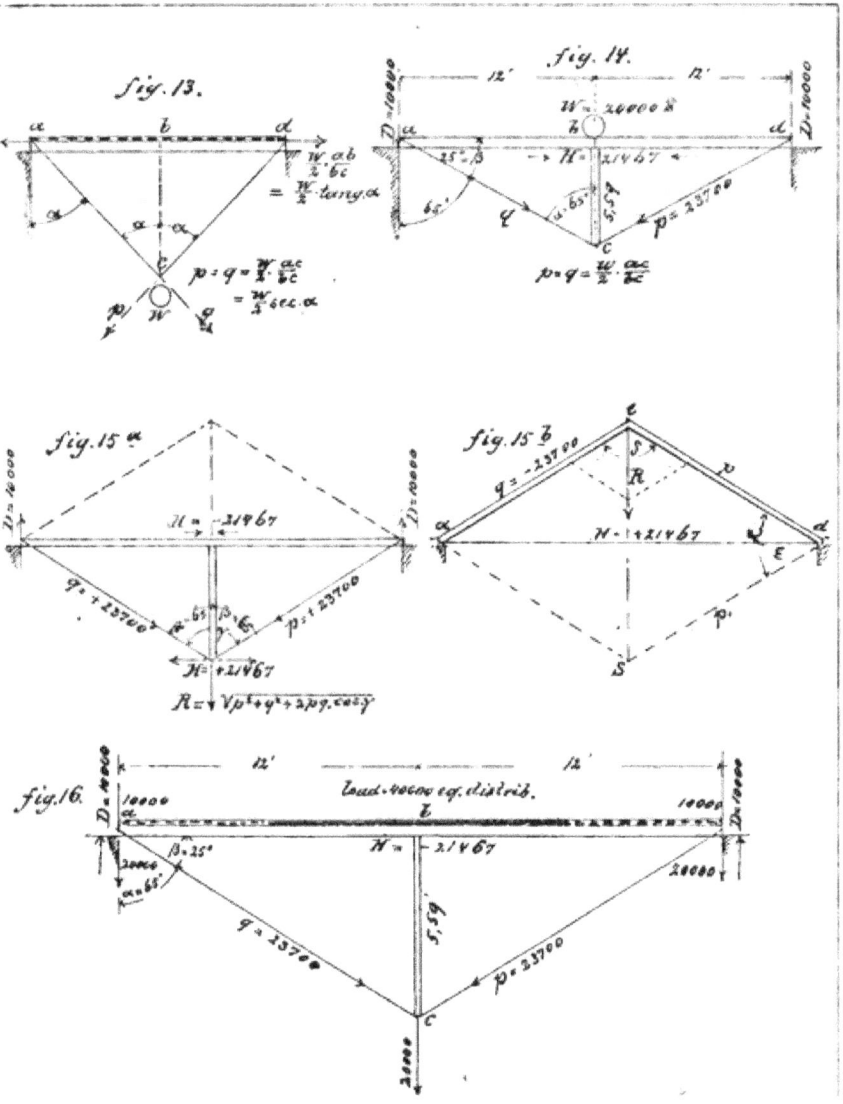

and (Fig. 14), the horizontal strain,

$$H = \frac{W}{2} \cdot \frac{12}{6} = \frac{20000}{2} \times \frac{12}{6}, \quad \text{or } H = 20000,$$

thus being in this case the same as the weight, W, at the centre.

The horizontal strain (thrust) and the strain on the oblique rods increase with the angle α, thus being ∞ for an angle $\alpha = 90°$.

Suspension Truss Bridge.

We find a combination of trussing in the well-known "Suspension Truss" bridges, the principles for calculation of the strains being contained in the preceding.

The *Bollmann truss* forms a continuous system of independent trusses, in number equal to the number of vertical posts combined to a common top chord (stretcher).

Plate 4,] By Fig. 17 the dimensions of such a truss may be
Fig. 17.] represented.

When for a single-track railroad-bridge the assumed load, including the weight of structure $= 1\frac{1}{2}$ tons $= 3360$ lbs. per lineal foot, or for one rib (single truss) $= 1680$ lbs. per lineal foot—*i. e.*, for the given dimensions $12 \times 1680 = 20160$ lbs. on each post, for which may be said in round figures 20000 lbs. $= W$, for calculation, then in Fig. 17, according to Fig. 9, the tension in the first rod nearest the abutment,

Strain No. 1, rod $= W \cdot \dfrac{bF}{AF} \cdot \dfrac{Ak}{bk} = 20000 \times \dfrac{7}{8} \times \dfrac{15.6}{10} = 27300$ lbs.,

the section of which for a value of iron $= 10000$ lbs. per square inch (five to six times security) $= 2.73$ square inches; thus, when two rods are applied, the size of each rod $= 1 \times 1\frac{3}{8}''$.

Strain No. 2, rod $= W \cdot \dfrac{cF}{AF} \cdot \dfrac{Al}{cl} = 20000 \times \dfrac{6}{8} \times \dfrac{26}{10} = 39000$ lbs.

Section $= 3.9$ sq. in., or 2 rods, each $1 \times 2''$.

Strain No. 3, rod $= W \cdot \dfrac{dF}{AF} \cdot \dfrac{Am}{dm} = 20000 \times \dfrac{5}{8} \times \dfrac{37.3}{10} = 46625$ lbs.

Sect. $= 4.66$, or 2 rods, each $1 \times 2\frac{3}{8}''$.

Strain No. 4, rod $= W \cdot \dfrac{EF}{AF} \cdot \dfrac{AN}{EN} = 20000 \times \dfrac{4}{8} \times \dfrac{49}{10} = 49000$ lbs.

Sect. $= 4.90$, or 2 rods, each $1 \times 2\frac{1}{2}''$.

Strain No. 5, rod $= W \cdot \dfrac{fF}{AF} \cdot \dfrac{Ao}{fo} = 20000 \times \dfrac{3}{8} \times \dfrac{60,8}{10} = 45600$ lbs.

Sect. $= 4,56$, or 2 rods, each $1 \times 2\frac{1}{4}''$.

Strain No. 6, rod $= W \cdot \dfrac{gF}{AF} \cdot \dfrac{Ap}{gp} = 20000 \times \dfrac{2}{8} \times \dfrac{72,6}{10} = 36300$ lbs.

Sect. $= 3,63$, or 2 rods, each $1 \times 1\frac{7}{8}''$.

Strain No. 7, rod $= W \cdot \dfrac{hF}{AF} \cdot \dfrac{Aq}{hq} = 20000 \times \dfrac{1}{8} \times \dfrac{84,6}{10} = 21150$ lbs.

Sect. $= 2,11$, or 2 rods, each $1 \times 1\frac{1}{4}''$.

When for a partial load at a certain panel the exertion of a pair of suspenders is greater than for a distributed load in calculation, those rods would be strained more than to one-fifth or one-sixth of their ultimate strength. So, when a locomotive of 84000 lbs. weight rests at a certain panel on a wheel-base of 12 feet, to each of the four supporting posts would be transmitted one-fourth of its weight $= 21000$ lbs.—this differing very little from the calculation in the example. Additional rods (panel-rods) are applied, sustaining the main suspenders and at the same time the top chord, transmitting and distributing the weight on the posts, these being always in a state of compression equal to the weight on the post.

Without the panel-rods for an over-grade bridge (through bridge) there would be in the post no further compression than that produced by the weight of the top chord and appendages, leaving for a strong cambered truss (B, Sect. II.), in case of a partial load, the possibility of raising.

The following is the strain in panel-rods according to Fig. 8*:

Strain $= \dfrac{W}{2} \cdot \dfrac{Er}{fe} = \dfrac{20000}{2} \times \dfrac{16}{10,5} = 15228$ lbs.

Sect. $= 1,52$, or 1 rod $= 1 \times 1\frac{1}{2}''$.

The influence of temperature upon the single systems of main suspenders (their length being different) is regulated by a link connection.

For the compressive strain in the top chord the rule for a girder, sustained at both ends and charged with an equally-distributed load, may be applied (see at the close of this chapter), then,

SUSPENSION TRUSS BRIDGE.

$$\frac{20000 \times \frac{7}{2} \times \frac{9.3}{4}}{10} = 168000.*$$

The compression in the top chord is the same all over.

For the result we have as momentum one-half of the entire weight on posts at one-fourth the length of truss as leverage,

or
$$\text{Mom.} = \frac{Q}{2} \times \frac{L}{4},$$

which, when divided by the depth of truss, gives the compression = 168000 lbs. as before.

For a single load, P, at the centre would be

$$\text{Mom.} = \frac{P}{2} \times \frac{L}{2},$$

which, when divided by the depth of truss, gives for the compression twice as much, or 336000 lbs.; but for an addition of the results of each single truss with its single load, according to x and x_1 in Fig. 9, it would be

$$\text{Mom.} = \frac{P}{2} \times \frac{3L}{8},$$

this being one-half of the result for a single load, P, at the centre, added to one-half of the result for an equally-distributed load, Q.

18.] The *Fink truss* (Fig. 18) is different in principle.

Whilst in the Bollmann system there are as many independent trusses as there are posts, in the Fink all the trusses are dependent on each other and transfer the load toward the centre.

The centre post (king-post) has to sustain the compression of one-half of the entire load on the truss, including one-fourth of the weight of the rib, the main suspenders (tie-rods) depending again, as before, on the depth of the truss.

---

* For a simple compressive strain an area of section of stretcher = 9 square inches would be sufficient (18000 lbs.—safe load for cast iron)—the actual dimensions to be taken by Hodgkinson's formula on the strength of hollow cast-iron pillars:

$$W = \text{breaking weight in tons} = 44.3 \cdot \frac{D^{3.6} - d^{3.6}}{L^{1.7}};$$

therefore, when for a pillar, the external diameter, D, in inches, and the length, L, in feet, are known; and for six times security, with the weight, W, multiplied by 6, we can define the internal diameter, d, consequently the thickness of metal.

3

The calculation of an example in its simplicity will give the best explanation.

Taking the same dimensions and the same load as in the calculation for the preceding (Fig. 17), according to Fig. 8^b we have,

$$\text{Strain in } A.N \text{ or } IN = \frac{W}{2} \times \frac{AN}{EN};$$

i.e., $\quad \dfrac{80000}{2} \times \dfrac{49}{10} = 196000$ lbs.,

the section of which for a value of iron $= 10000$ lbs. per square inch $= 19,6$ square inches. Thus, when two rods are applied, the size of each $= 2 \times 5$ inches.

$$\text{Strain in } kc, Ak \text{ or } cm = \frac{20000}{2} \times \frac{13}{5} = 26000 \text{ lbs.}$$

Section of a single rod $= 1 \times 2\frac{1}{2}$ in. (full).

$$\text{Strain in } Al \text{ or } lE = \frac{40000}{2} \cdot \frac{26}{10} = 52000 \text{ lbs.}$$

Section of a single rod $= 2 \times 2\frac{1}{2}$ in., or 1×5 in.

For a single locomotive (weight 84000 lbs.), resting at cd on a wheel-base of 12 feet, the vertical force for one post at c or $d =$

$$\frac{42000}{2} = 21000 \text{ lbs.} = W;$$

then, $\quad$ Strain in tie-rods $= \dfrac{21000}{2} \times \dfrac{13}{5} = 27300$ lbs.;

so the size of rod kc, Ak or cm should be corrected to $1 \times 2\frac{1}{2}$ in.

For the compressive strain on the top chord (stretcher), according to Fig. 14,

$$H = \frac{W}{2} \cdot \frac{AE}{EN},$$

and for a vertical strain $W = 80000$ lbs. in the centre post, as before mentioned,

$$H = \frac{80000}{2} \times \frac{48}{10} = 192000 \text{ lbs. (compr.).}$$

In this truss, as in the Bollmann, the compression in the top chord is the same all over.

For this truss, when applied for an over-grade railroad-bridge, a safe longitudinal connection (bracing or tying) will be essential on account of the variable load.

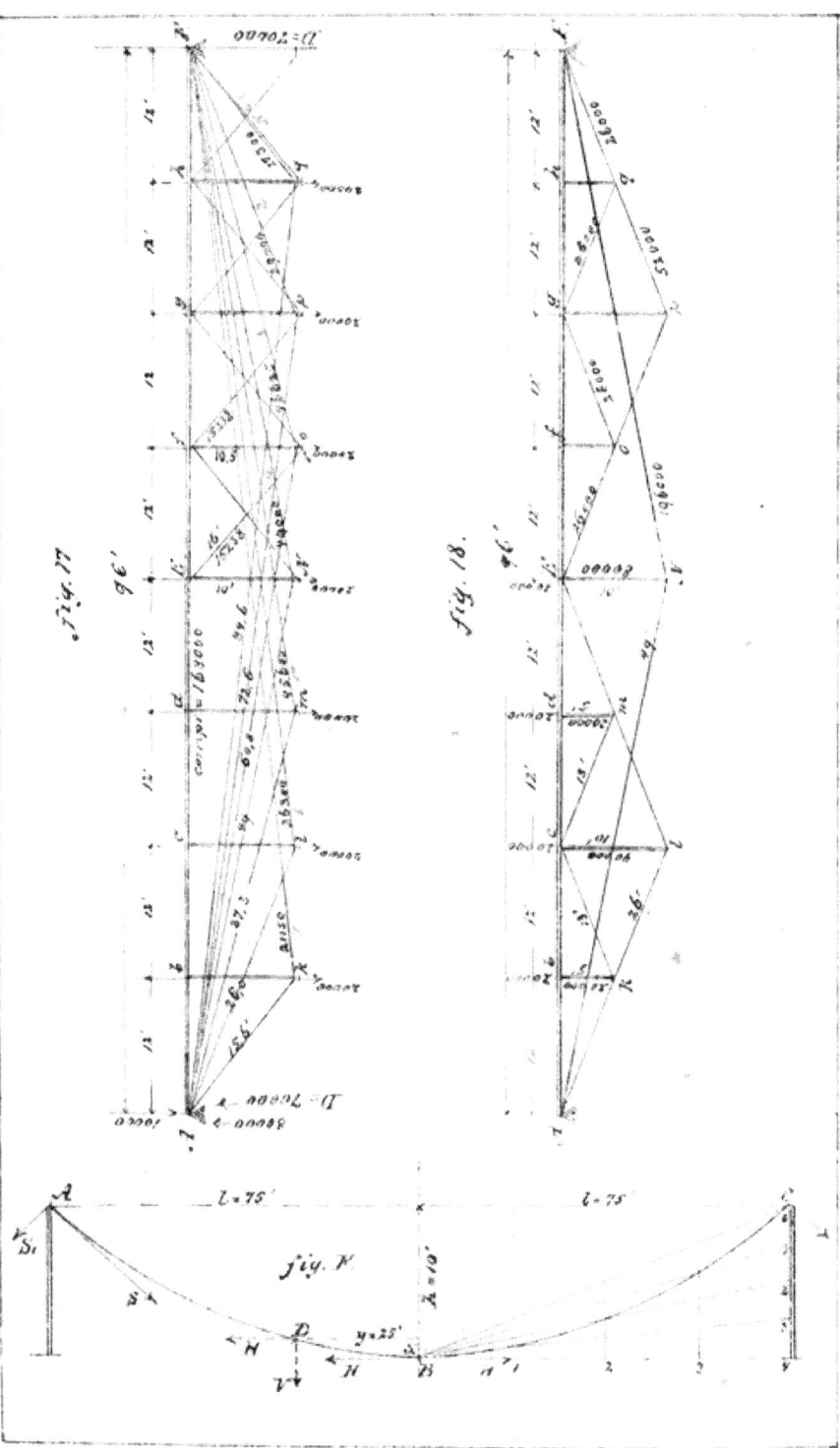

To the same category of bridges belongs the

SUSPENSION BRIDGE,

though, in regard to mathematics, very different.

The curve formed by a chain or cable lies between the parabola and the catenary, and is very nearly an ellipse. The curve in a loaded state approaches the parabola; in an unloaded state, the catenary. (Weisbach, vol. II.)

In the following example the curve may be considered a parabola or the bridge in its loaded state.

The thesis is—

The vertical force at every point of the chain equals the weight on the chain from the point in consideration unto the vertex.

Plate 4,] So, when y in Fig. F = 25 feet and the length of
Fig. F.] bridge = 150 feet,

Width = 4 feet;

load, 50 lbs. per square foot, or 200 lbs. per lineal foot;

maximum load = 200×150 = 30000 lbs.;

the vertical force at $D = 200 \times 25 =$ 5000 lbs.

Further, *the horizontal force at every point of the chain is equal, and therefore equal to the horizontal strain in the vertex.*

Thus, when by p is represented the weight *pro unit* of horizontal projection = 200 lbs., for our example,

I. $H = \dfrac{pl^2}{2h}$, which is at the same time the horizontal force in A and C, to overturn the towers, and amounting here to $\dfrac{200 \times 75^2}{20} =$ 56250 lbs.

II. $S = \dfrac{pl}{2h} \cdot \sqrt{l^2 + 4h^2} = \dfrac{15000}{20} \times 77{,}6 =$ 58500 lbs.

III. $y^2 = \dfrac{l^2}{h} \cdot x.$

The length, L, of chain results by the formula,

$$L = 2l + \dfrac{4}{3} \times \dfrac{h^2}{l} = 151{,}75.*$$

---

* For more specifications I ought to refer to Weisbach and other authors.

For a trussed system with two posts (queen-posts), between supports 36 feet, and an assumed load $= 30000$ lbs., equally distributed, the distribution of forces on the bearings is denoted in Fig. 19, and for the calculation of strain, when compared with Fig. 10, we find for

[Plate 5, Fig. 19.]

$$Y = W \cdot \frac{ad}{bd};$$

$$Y = 10000 \times \frac{13,24}{5,59} = 23700 \text{ (approx.)};$$

and for
$$Y_1 = W \cdot \frac{ab}{bd};$$

$$Y_1 = 10000 \times \frac{12}{5,59} = 21467.$$

The compression on the vertical posts $= 10000$ lbs.; the vertical pressure on supports $= 15000$ lbs.; and, reduced by those 5000 lbs. directly sustained (comp. Fig. 16), the reactive force of supports, signified by $D = 10000$ lbs.

Upon this structure, the *Method of Moments* being applied (see Preface), we suppose a section separated from the original by a cut, *st*.

Considering the forces acting upon such a section, we form the equation of equilibrium for a suitable point of rotation, by solution of which we find the strain in the member in question; and observe the rule, that, *for a strain, Y, the point of rotation ought to be chosen in the intersection of x and z, making their lever $= 0$, when these are the members of the structure, separated by a cut, st, likewise as Y.*

[20.] But when by *st* only one member, excepting Y, is separated, we lay the point of rotation on the next joint, as, per example, for Y in b, Fig. 20.

$$Y \cdot 5{,}07 = 10000 \times 12,$$

or, I., $\quad 0 = -Y \cdot 5{,}07 + 10000 \times 12$ (rotation round b);

$$Y = \frac{10000 \times 12}{5{,}07} = +23700;$$

and because in the following the form of equation always will be kept similar to I.; the forces in their aim to turn to the left, like Y round b, will be signified by $-$, the same as a compressive strain; and the forces to the right, like the hands on a watch, or D round b, will be signified by $+$, the same as a tensile strain.

21.] According to this we have for Z, Fig. 21,

$$0 = Z \cdot 5{,}59 + D \cdot 12 \text{ (rot. } r \cdot d),$$
$$Z = \frac{10000 \times 12}{5{,}59} = 21467,$$

and for Y_1, Fig. 22,

22.] $\quad 0 = - Y_1 \cdot 5{,}59 + D \cdot 12 \text{ (rot. } r \cdot b);$

$$Y_1 = \frac{10000 \times 12}{5{,}59} = 21467.$$

When a diagonal, s, sustains the parallelogram, so that by a cut, st, three members, Z_1, Y_1 and s are separated, we have for the definition of s the point of rotation, as before mentioned, in the intersection of Z_1 and Y_1; but Z_1 and Y_1 in their direction are parallel, and therefore without intersection at all. In this case (the same as for diagonals in girders with parallel top and bottom flanges) we

23.] suppose a point of rotation, O, at any distance in the direction (axis) x, and find thus from Fig. 23, where $x = \infty$, or infinite—i.e., the lever of all forces, acting in a vertical direction upon the section $= x$.

$$0 = -s \cdot \infty \sin \varphi - D \cdot x + p \cdot \infty \text{ (rot. } r \cdot O);$$

or, because ∞ is a factor of each part,

$$0 = -s \cdot \sin \varphi - 10000 + 10000.$$

24.] In this equation, $s \cdot \sin \varphi$ (comp. Fig. 56 on semi-girders) is the vertical component of s (Fig. 24), and acts right-angled to the axis, x, like D and p.

The angle φ for our example $= 25°$;

$$\sin \varphi = 0{,}422 \text{ (see table)},$$

and therefore $\quad 0 = -s \times 0{,}422 - 10000 + 10000,$

or $\quad s = 0,$

showing that, as in Fig. 19, the diagonal, s, is without any strain and only useful for preventing dislocation. (Compare parabolic

25.] girders, to which this case is similar, because a parabola can be constructed through the sustaining points a, d, c and c, and therefore differs in this from Figs. 25 and 63.)

26.] For a reversed structure (Fig. 26) the strains will be the same, but of reversed signs.

In the following calculations of roofs and bridges it will be shown that the *Method of Moments* is thoroughly applicable, leading directly and in the most comprehensive manner to distinct results; but for a preliminary estimate of strain in the top and bottom chords at the centre of the structure the most simple way to define this strain may be stated by the following:

27,]
28.] As the flanged girder in Fig. 27, charged with an equally-distributed load, Q, will be exposed at its centre to the same exertion as the girder in Fig. 28, fixed at the centre, the moments will be for both, when $V =$ depth of girder and $l =$ length (the load being equally distributed).

$$\text{Mom} = \frac{Q}{2} \cdot \frac{l}{4} = V \cdot x \text{ (rot. } r \cdot o\text{)};$$

$$\frac{\frac{Q}{2} \cdot \frac{l}{4}}{V} = \frac{\frac{1}{8} Q \cdot l}{V} = \text{compression or tension in flanges};$$

so $Q = 30000$ lbs.; $V = 5{,}59'$; and $l = 36'$.

$$\text{Mom} = \frac{30000}{2} \times \frac{36}{4} = 135000 = V \cdot x,$$

and $\frac{135000}{5{,}59} = 24150$ lbs., the strain in the chords.

It is to be observed that the result is rather too high when applied upon a truss with few panels, as in Fig. 26, on account of the reactive pressure of the support, diminished by the partition of the direct load on this place.

On account of its being a very convenient method we recommend it; and in the following bridge skeletons we refer to it very frequently. (See D, Example I., and Sect II., note.)

29.] For a girder charged with a single weight at the centre (Fig. 29), we make a comparison with Fig. 30, and find for both

$$\text{Mom} = \frac{P}{2} \cdot \frac{l}{2} = V \cdot x = \tfrac{1}{4} P \cdot l,$$

30.] and $\quad \frac{\frac{1}{4} P \cdot l}{V} = \text{horizontal strain};$

i.e., compression or tension in the top or bottom chords, the formation of moments for a point of rotation, o, being very comprehensive.

When $\quad\quad P = 15000$ lbs.,

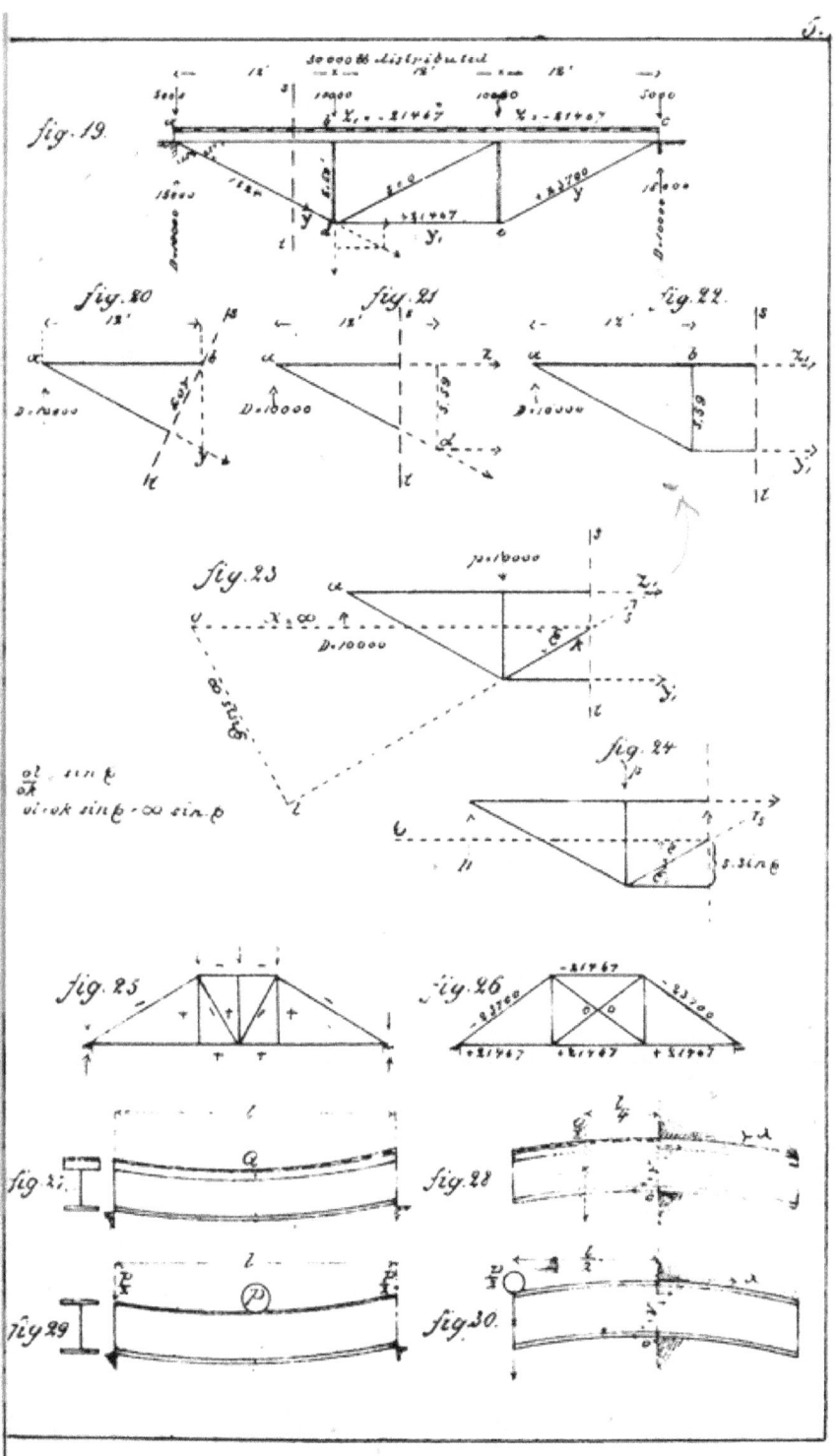

4

but $V = 5{,}59$, and $l = 36$ (as before),

so the $$\text{Mom} = \frac{15000}{2} \times \frac{36}{2} = 135000,$$

and $$\frac{\frac{15000}{2} \times \frac{36}{2}}{5{,}59} = 24150 \text{ lbs., the strain in flanges};$$

showing that, for the same exertion, a beam loaded with a single weight, P, at the centre can bear only one-half of an equally-distributed load, Q.

[Plates 3, 4 and 5—embracing Figs. 13 to 30.]

B. ROOF CONSTRUCTION.

For small and not complicated roofs, experience is a common and in general, also, a sufficient guide. But experience is very limited, and not every constructor has opportunity and time to acquire it.

The true and acceptable guide for a safe practice will always be the calculation; and since, especially for complicated and more extensive combinations, the application of mechanical science has become unavoidable, the following compendium, leading from the simplest to the most complicated structures, will give for almost every purpose sufficient information.

[Plate 6, Fig. 31ª.] By construction, when EG represents the weight at the centre of gravity, G (Fig. 31ª), the body will be at rest when the plane DF is right-angled to the line DE; CE being horizontal—i. e., right-angled to the vertical line EG.

Let EG be an assumed length, then in the parallelogram of forces the intensity of EK and EH is measured in proportion by the same rule.

[31ᵇ.] For an angle, $\beta = 25°$ of a rafter with the horizontal line (Fig. 31ᵇ—similar to Fig. 16 reversed) leaning with the top end against a wall, the heel at A being morticed; we then have

$$\frac{cb}{ca} = \sin \beta = \frac{5{,}59}{13{,}24} = 0{,}422;$$

$$\sin^2 \beta = 0{,}17;$$

$$\sin 2\beta = \sin 50° = 0{,}766;$$

$$\frac{ab}{ac} = \cos \beta = \frac{12}{13,24} = 0,906;$$

$$\cos^2 \beta = 0,82.$$

When W is an equally-distributed load of 20000 lbs., then

$$H = H_1 = \frac{W}{2} \times \frac{d}{h} = \frac{20000}{2} \times \frac{12}{5,59} = 21467 \text{ lbs.}$$

The vertical force, V, at the top of the rafter $= 0$, and the vertical pressure V_2 of the heel $= 20000$ lbs., or equal to the entire load.

Also the vertical force, V_1 at the centre $= 20000$ lbs.

The pressure in the rafter itself (compression) $=$

$$\frac{W}{2} \times \frac{l}{h} = 10000 \times \frac{13,24}{5,59} = 23685 \text{ (23700 lbs.)}.$$

The entire pressure, R, of rafter toward the support,

$$R = \sqrt{V_2^2 + H_1^2} = 29300 \text{ lbs.}$$

Its direction can be constructed in making $K = 2h = 11,18$ feet, H_1 and V_2 forming the sides of the parallelogram.

When $P = 10000$ lbs., a single weight at centre of rafter,

$$H = H_1 = \frac{P}{2} \cdot \frac{d}{h} = 10733;$$

$$V = o,$$

and $\qquad V_2 = V_1 = 10000$ lbs.

[Fig. 31°.] When in Fig. 31°, by FM the weight of the body $ABCD$ is represented, then FN, the force toward the wall, results in the horizontal and vertical forces CH and CV.

FL, the force acting perpendicular to the plane BI, in the direction BF.

G, the centre of gravity.

GF, vertical.

$CF \perp CB$, or $\angle FCB = 90°$.

$BI \perp FB$, then BI is the direction of the plane required to make the body at rest.

Also $\qquad FN = P = \frac{s}{l} G \cos \alpha;$

$$H = P \sin \alpha = \tfrac{1}{2} G \sin 2\alpha.$$

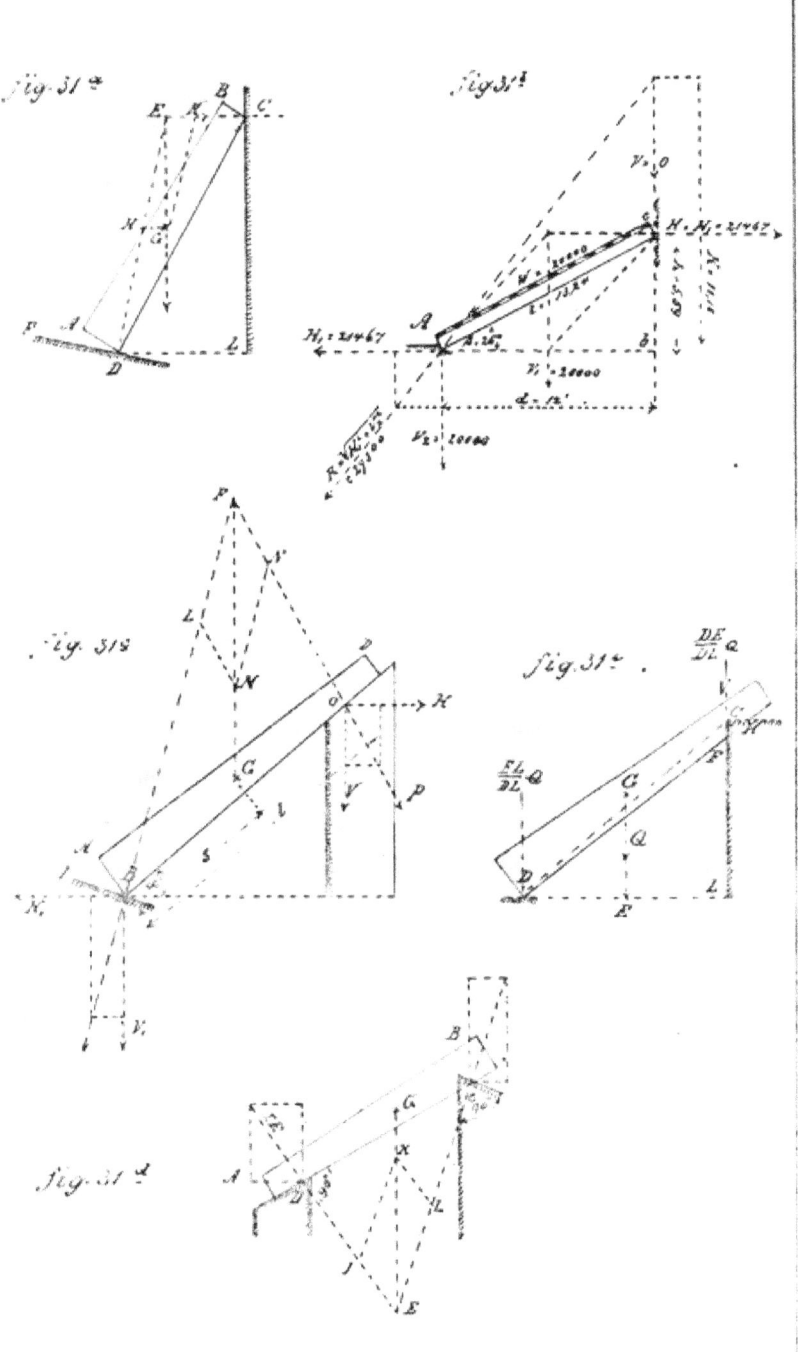

B. ROOF CONSTRUCTION.

31^a.] Let the sloping body $ABCD$ (Fig. 31^a) be supported by a wall at its lower end, D, which coincides with the surface of the body;

Let G be again the centre of gravity;

It is required to cut a notch out of the body at the upper end, C, so that it may rest upon the top of a wall which is made to fit the notch.

Make GE vertical;

From D draw $DE \perp$ to CD;

Join EC, and draw CF at right angles to it; then the notch at C being cut, the body $ABCD$ will be at rest.

31^e.] A body, $ABCD$, resting on supports (Fig. 31^e), will only produce the vertical strains $\dfrac{DE}{DL} \cdot Q$ and $\dfrac{EL}{DL} \cdot Q$ at the supports.

Plate 7, Fig. 32.] For Figs. 32, 33 and 34 we have, again,

$$\frac{bc}{ac} = \sin \beta ; \quad \frac{ab}{ac} = \cos \beta ; \quad \frac{bc}{ab} = \tan \beta ; \quad \frac{ab}{bc} = \cotg \beta ;$$

$$\sin 2\beta = \sin 50° = 0{,}766 ;$$

and when upon each rafter $W = 20000$ lbs., equally distributed, we have for Fig. 32,

$$H = H_1 = \frac{W}{2} \cdot \frac{d}{h} = \frac{W}{2} \cdot \cotg \beta = 21467 \text{ lbs.}$$

The compression in the rafter $= \dfrac{W}{2} \cdot \dfrac{l}{h} = 23685$ lbs.,

and, again, $\quad R = \sqrt{H_1^2 + V_2^2} = 29300$ lbs.

For an angle $\beta = 26° \ 34'$; the horizontal thrust will $= 20000$ lbs.—i. e., the same as the entire weight. (Comp. Fig. 16.)

33.] For Fig. 33 is, as the rafter in a vertical direction, sustained on the top,

$$H = H_1 = \frac{W}{4} \cdot \sin 2\beta = 5000 \times 0{,}766 \times 3830 ;$$

$$V = \frac{W}{2} \cdot \cos^2 \beta = 10000 \times 0{,}82 = 8200 ;$$

$$V_2 = \frac{W}{2} (1 + \sin^2 \beta) = 10000 (1 + 0{,}17) + 11700 ;$$

and the compression in the rafter,

$$\frac{W}{2}\cdot \sin \beta = 10000 \times 0{,}422 = 4220 \text{ lbs.}$$

For Fig. 34, when the rafter is sustained at the top by a vertical post,

$$H = H_1 = \frac{W}{4}\cdot \sin 2\beta = 3830;$$

34.] $V = W \cos^2 \beta = 20000 \times 0{,}82 = 16400;$

$$V_2 = \frac{W}{2}(1 + \sin^2 \beta) = 10000 (1 + 0{,}17) = 11700;$$

and the compression in the rafter =

$$\frac{W}{2}\cdot \sin \beta = 4220 \text{ lbs.}$$

In the cases in Figs. 33 and 34 the post relieves the tie-rod or (as here, in the absence of a tie) the wall from a part of the thrust of the rafters, and compared with the truss in the preceding we see that the king-post acts in a different way.

The different combinations of this single hanging-and-thrust construction may be disregarded, and by the following calculations the method of moments thoroughly applied.

35.] A rafter being constructed in this way (Fig. 35), without connection between the point e and the horizontal tie-rod aa, there are two different systems, that of a trussed beam, ace, similar to Fig. 16, and that of a single triangle like Fig. 32, and it is as there for the same angle, β, and a load = 20000 lbs. upon each rafter; the horizontal thrust, H, at the top, as also the tension in the horizontal tie-rod = 21467 lbs. The trussed rafter is understood to be calculated like Fig. 16 in combination with Fig. 31.

36.] When the weight of structure, snow and wind-pressure upon a roof (Fig. 36), for one square foot of horizontal projection, in all = 50 lbs., and the width between the supports or AC = 40 feet, the distance of rafters = 10 feet, then 20000 lbs. is the entire load, or 10000 lbs. upon each rafter, supported at A, B, C, D and E.

The pressure of one-half of the weight, or 10000 lbs., on the supports is counteracted by the direct load of 2500 lbs. It is therefore the reacting force, D, only 7500 lbs. (See Howe Truss, Sect. I., D.)

For a section separated by st we can define at once the strains in

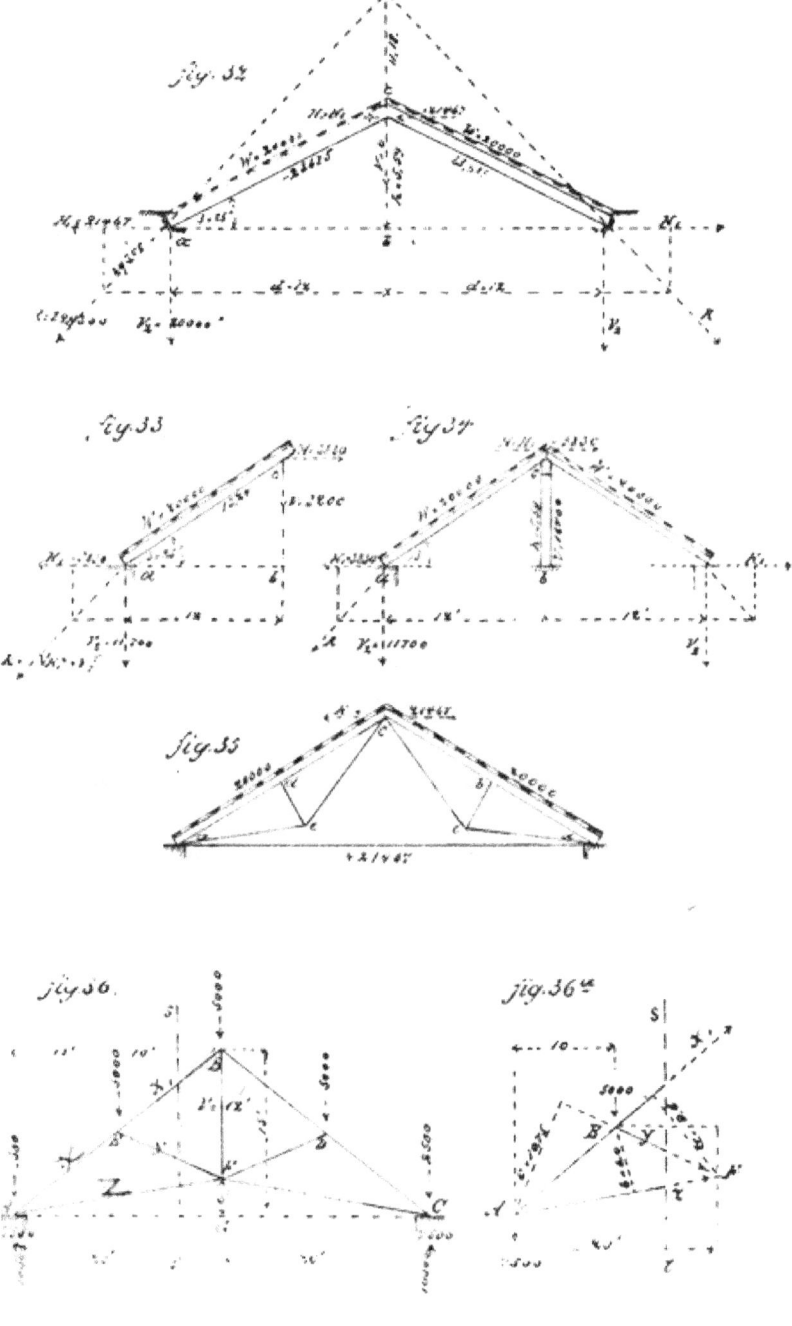

z, y and x_1, so for x_1, considering the forces acting upon this section and the point of rotation in the intersection of y and z or F. (Fig. 36ᵃ.)

36ᵃ.]
$$0 = x_1 . a + D . 20 - 5000 \times 10.$$

(Comp. Fig. 20, Equat. I.)

The arm or lever, a, can be measured near enough from a skeleton $= 9,6'$.

Besides, for the calculation of a we have for the angle ABG (Fig. 36),

$$AG = GB . \tan < ABG,$$

or $\qquad \tan . ABG = \dfrac{AG}{GB} = \dfrac{20}{15} = \dfrac{4}{3} = 1,333;$

i. e., $\qquad < ABG = 53° 7'$, and $\sin 53° 7' = 0,8;$

$$a = BF \sin < ABG = 12 \times 0,8 = 9,6;$$

therefore $\quad 0 = x_1 . 9,6 + 7500 \times 20 - 5000 \times 10;$

$$x_1 = -\frac{100000}{9,6} = -10412 \text{ lbs.}$$

For z (rot. r . E, Fig. 36ᵃ),

$$0 = -z . 6 + 7500 \times 10,$$

or $\qquad z = +\dfrac{75000}{6} = +12500;$

and for y (rot. r . A, Fig. 36ᵃ),

$$0 = y . c + 5000 . 10;$$
$$0 = y . 10,75 + 50000,$$

or $\qquad y = -\dfrac{50000}{10,75} = -4650.$

Plate 8, Fig. 36ᵇ.] For the strain in x (rot. r . F, Fig. 36ᵇ) is,

$$0 = x . 9,6 + 7500 \times 20,$$

or $\qquad x = -\dfrac{150000}{9,6} = -15625.$

In regard to the vertical V, we use for its definition the strain of the joining brace, $x_1 = -10412$, and make it a curved line; then we have, for a rot. r . D (Fig. 36ᵇ),

$$0 = -V . 10 - (-10412) . 10,9,$$

or $\qquad 0 = -V . 10 + 113490;$

36 THE THEORY OF STRAINS.

i.e.,
$$V = + \frac{113490}{10} = + 11349.$$

37.] The results are combined in Fig. 37.

38.] For Fig. 38, the entire load (equally distributed) again being 20000 lbs., the depth 15′, and between the supports 40′.
When here the cut *st* separates the line $x_1 y_1 z_1$, we have for x_1 (rot. in the intersection of $y_1 z_1$, or F, Fig. 39),

$$0 = x_1 . 9,1 - 5000 \times 5\tfrac{1}{2} + 7500 \times 15\tfrac{1}{2};$$

39.]
$$x_1 = -\frac{88750}{9,1} = 9752.$$

For y_1 (rot. *r* . *A*), $0 = - y_1 . 15 + 5000 \times 10,$

or
$$y_1 = + \frac{50000}{15} = + 3,333;$$

and for z_1 (rot. *r* . *C*),

$$0 = - z_1 . 15 - 5000 \times 10 + 7500 \times 20;$$
$$z_1 = + 6,666.$$

40.] For a section, *st*, through *x* and *z*, we have for *x* (Fig. 40),

$$0 = x . 9,1 + 7500 \times 15\tfrac{1}{2} \text{ (rot. } r . F\text{)};$$

$$x = - \frac{116250}{9,1} = - 12774;$$

and for *z* (rot. *r* . *D*),

$$0 = - z . 7,4 + 7500 \times 10;$$

or
$$z = \frac{75000}{7,4} = 10135.$$

For *y* (rot. *r* . *A*) is, $0 = y . 12,5 + 5000 \times 10;$

$$y = - \frac{50000}{12,5} = - 4000.$$

41.] The results combined in Fig. 41.

42.] When the figure before is changed in the depth, like Fig. 42, we have the following equation:

$$0 = x_1 . 5,8 - 5000 \times 3,25 + 7500 \times 13,25 \text{ (rot. } r . F, \text{Fig. 43)};$$

$$x_1 = - \frac{83125}{5,8} = - 14332.$$

43.] For the definition of y_1, the intersection of x_1 and z_1 will be in G, and it is for C as rotation,

$$0 = - y_1 . 8,25 + 5000 \times 6 + 7500 \times 4; \quad \text{(Fig. 43.)}$$

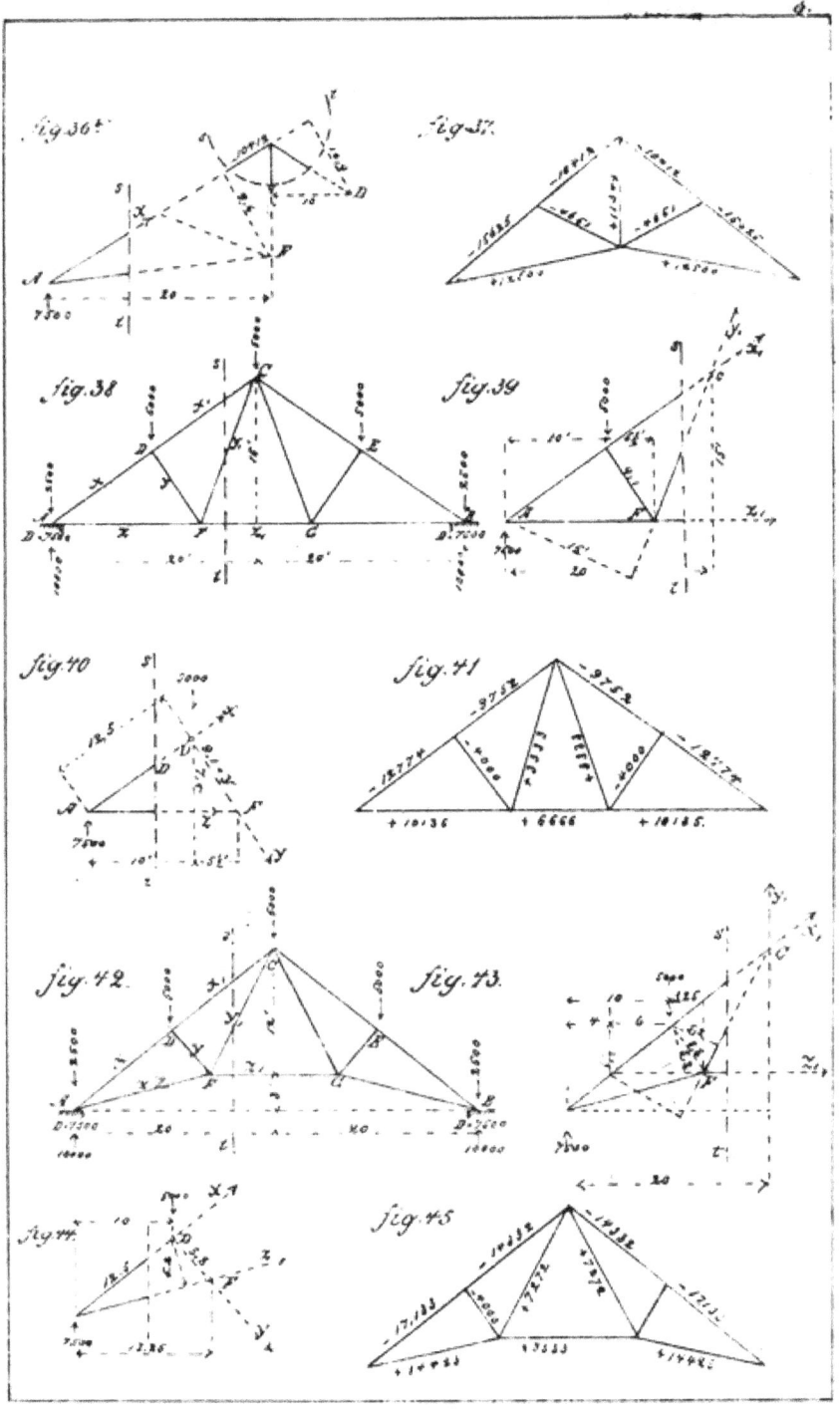

B. ROOF CONSTRUCTION.

$$y_1 = + \frac{60000}{8{,}25} = 7272.$$

For z_1 (rot. $r \cdot C$) we have

$$0 = -z_1 \cdot 12 - 5000 \times 10 + 7500 \times 20;$$

$$z_1 = + \frac{100000}{12} = +8333.$$

44.] For a section, st, through x and z (Fig. 44), it is

$$0 = x \cdot 5{,}8 + 7500 \times 13{,}25 \text{ (rot. } r \cdot F);$$

$$x = -\frac{993750}{5{,}8} = -17133;$$

and

$$0 = -z \cdot 5{,}2 + 7500 \times 10 \text{ (rot. } r \cdot D);$$

$$z = \frac{75000}{5{,}2} = +14423;$$

and for y,

$$0 = y \cdot 12{,}5 + 5000 \times 10 \text{ (rot. } r \cdot A);$$

$$y = -\frac{50000}{12{,}5} = -4000.$$

45.] The results combined in Fig. 45.

Plate 9,] For the definition of X in Fig. 46, the point of rotation
Fig. 46.] in E, or in the intersection of Y and Z, will be from
Fig. 47.

47.]

$$0 = X \cdot x - P \cdot CE + D \cdot AE;$$

or

$$X = \frac{P \cdot CE - D \cdot AE}{x}.$$

For Y we choose A, or the intersection of X and Z, as the point of rotation, and the equation will be

$$0 = -Y \cdot y + P \cdot AC + Q \cdot AE,$$

or

$$Y = \frac{P \cdot AC + Q \cdot AE}{y},$$

and in the same way for Z, rot. $r \cdot H$.

$$0 = -Z \cdot z - Q \cdot EL - P \cdot CL + D \cdot AL,$$

or

$$Z = \frac{-Q \cdot EL - P \cdot CL + D \cdot AL}{z}.$$

It will not be necessary to show, by repetition of the foregoing, the equations for the other parts of the structure.

48.] In more complicated systems (Fig. 48), it may happen that by a cut, st (which can be made curved as well as straight), different braces or rods are spared, like FG, DG and DE.

In this case it is possible to come to a direct result when st only can be laid so that all the braces or rods cut by st meet at one point, except that one whose strain is in question.

49.] So for the strain V in FG (rot. r. H, Fig. 49),
$$0 = -V \cdot FH - R \cdot r;$$
$$V = -\frac{R \cdot r}{FH}.$$

50.] In the same manner the strain U in DG (rot. r. H, Fig. 50),
$$0 = U \cdot u - R \cdot r;$$
$$U = \frac{R \cdot r}{u};$$

thus we find also the strain in KT and LT.

51.] Being by the foregoing in possession of a value for U in DG, we find for the strain X in DF, Y in DE, and Z in CE the following equations from Fig. 51:
$$0 = X \cdot DE + U \cdot v - Q \cdot NO - P \cdot MO + W \cdot AO \text{ (rot. } r \cdot E);$$
$$0 = Y \cdot AD + U \cdot l + Q \cdot AN + P \cdot AM \text{ (rot. } r \cdot A);$$
$$0 = -Z \cdot z + W \cdot AN - P \cdot MN \text{ (rot. } r \cdot D);$$

each one enabling us to obtain a direct result for the strain in question.

Plate 10, Fig. 52.] For a roof (Fig. 52), the weight of which, 11,3 lbs. per square foot of its horizontal plan, may be calculated 20 lbs. for wind pressure and snow, making together 31,3 lbs. per square foot.

The distance of rafters being 15½ feet, the width, 100 feet, makes for each rafter 15½ × 100 × 31,3 = 48000 lbs. (approx.).

The load at each apex, therefore, will be $\frac{48000}{8}$ = 6000 lbs., the distribution of which is shown by the skeleton.

For the reactive force on the supports is again
$$D = 24000 - 3000, \text{ or } D = 21000 \text{ lbs.}$$

There are, in all, seven times 6000 lbs. acting downward, and twice 21000 lbs. acting vertically upward upon the system.

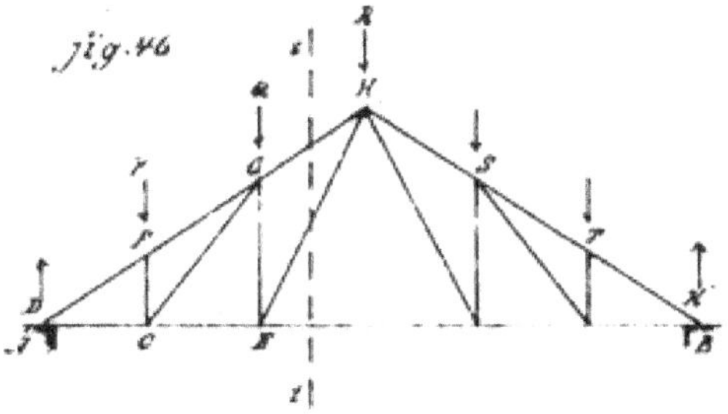

fig. 46

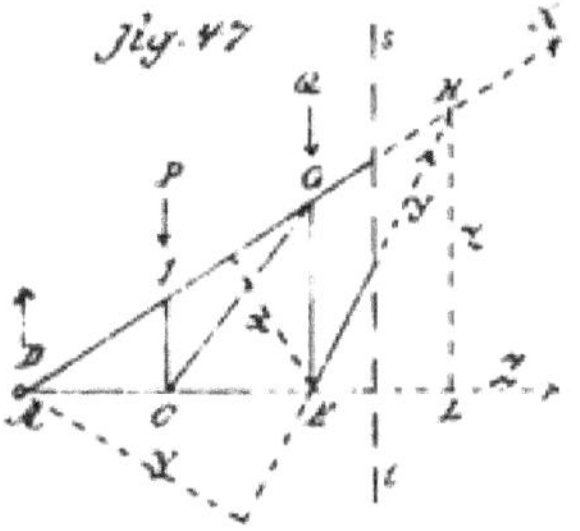

fig. 47

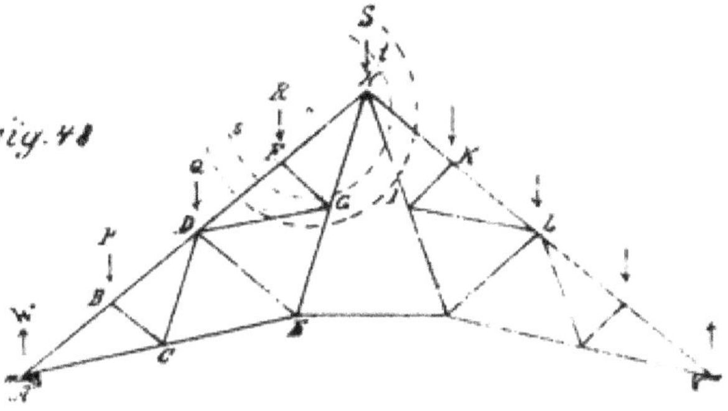

fig. 48

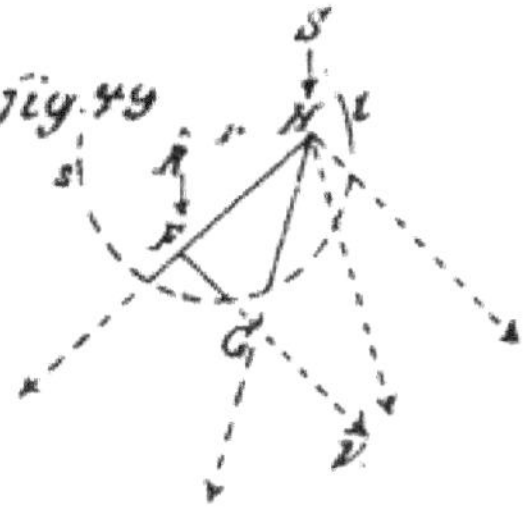

fig. 49

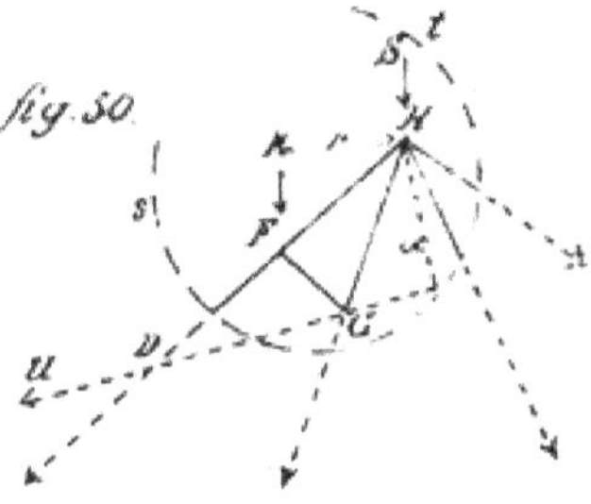

fig. 50

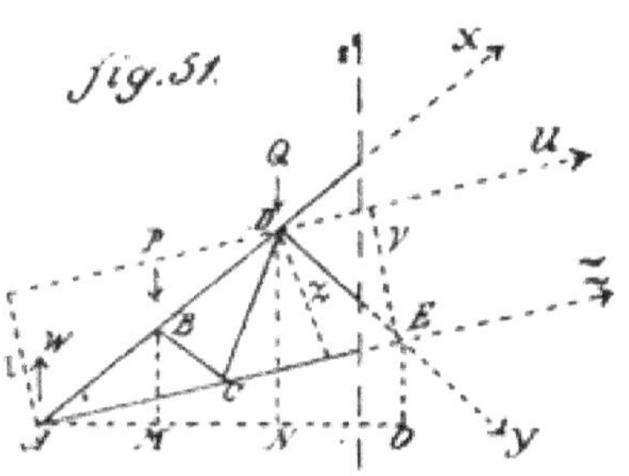

fig. 51

B. ROOF CONSTRUCTION.

53.] The section, A, s, t (Fig. 53), kept in equilibrium by the replaced forces, x, y and z, may be regarded first as a lever with the fulcrum at D; then the strain in x for the middle section is

$$0 = x \cdot 18,6 + 21000 \times 50 - 6000 \times 12,5 - 6000 \times 25 - 6000 \times 37,5,$$

or $\quad x = 32300$ lbs.;

and in y, when A is the point of rotation,

$$0 = y \cdot 38,4 + 6000 \times 12,5 + 6000 \times 25 + 6000 \times 37,5 \text{ (rot. } A);$$

$$y = -11700 \text{ lbs.,}$$

and

$$0 = -z \cdot 15 + 21000 \times 37,5 - 6000 \times 12,5 - 6000 \times 25 \text{ (rot. } r \cdot E);$$

$$z = +37500 \text{ lbs.}$$

54.] For V in Fig. 54 the rotation also round A is

$$0 = -V \cdot 37,5 + 6000 \times 12,5 + 6000 \times 25;$$

$$V = +6000 \text{ lbs.}$$

For the other members in Fig. 52,

$$0 = x_1 \cdot 13,9 + 21000 \times 37,5 - 6000 \times 12,5 - 6000 \times 25 \text{ (rot. } r \cdot F);$$

$$x_1 = -40400;$$

$$0 = y_1 \cdot 23,5 + 6000 \times 12,5 + 6000 \times 25 \text{ (rot. } r \cdot A);$$

$$y_1 = -9570;$$

$$0 = -z_1 \cdot 10 + 21000 \times 25 - 6000 \times 12,5 \text{ (rot. } r \cdot G);$$

$$z_1 = +45000;$$

$$0 = -V_1 \cdot 25 + 6000 \times 12,5 \text{ (rot. } r \cdot A);$$

$$V_1 = +3000;$$

$$0 = x_2 \cdot 9,3 + 21000 \times 25 - 6000 \times 12,5 \text{ (rot. } r \cdot H);$$

$$x_2 = -48400;$$

$$0 = y_2 \cdot 9,3 + 6000 \times 12,5 \text{ (rot. } r \cdot A);$$

$$y_2 = -8100;$$

$$0 = -z_2 \cdot 5 + 21000 \times 12,5 \text{ (rot. } r \cdot I);$$

$$z_2 = +52500.$$

For the strain in x_3 we choose a convenient point for rotation in the line z, per Example D, Fig. 55.

40 THE THEORY OF STRAINS.

55.] The equation in this case will be
$$0 = x_3 \cdot 18,6 + 21000 \times 50;$$
$$x_3 = -56500.$$

The only strain not directly deducible is U in the vertical line CD at the centre.

As in Fig. 36, we use the strain of the joining brace,
$$x = -32300 \text{ lbs.}$$

56.] For B as the point of rotation (Fig. 56), our equation is
$$0 = -U \cdot 50 - 6000 \times 50 - (-32300) \cdot 37,2;$$
$$U = 18000 \text{ lbs.}$$

57.] The results combined in Fig. 57.

58.] The weight and load of a roof (Fig. 58) being estimated, including wind-pressure and snow, to 50 lbs. per square foot of its horizontal plan, the distance of rafters being 12 feet, and the space between the walls 99 feet, which gives $50 \cdot 12 \cdot 99 = 59400$ lbs. for one rafter, or, in round figures, 60000 lbs.

The calculation of the top structure can be made as in the preceding example (Fig. 36).

58.] In the main construction are six supporting points, charged as in Fig. 58. The top structure transmits one-third of the entire load, or on each post 10000 lbs. to the apexes, ff.

Each wall has to bear 30000 lbs.; and after subtraction of the direct load the reactive force is 26666 lbs., or, by calculation,
$$D = \frac{6666 (11 + 22)}{99} + \frac{13333 (33 + 66)}{99} + \frac{6666 (77 + 88)}{99};$$
$$D = 26666 \text{ lbs.}$$

59.] For the strain x_3 we have in Fig. 59,
$$0 = -x_3 \cdot 21 - 13333 \times 16\tfrac{1}{2} - 6666 (27\tfrac{1}{2} + 38\tfrac{1}{2}) + D \cdot 49\tfrac{1}{2}$$
$$(\text{rot. } r \cdot h);$$
or, also,
$$0 = -x_3 \cdot 21 - 6666 (11 + 22) + 26666 \times 33 \;(\text{rot. } r \cdot f);$$
$$x_3 = +\frac{659967}{21} = +31427.$$

Further,
$$0 = Z_4 \cdot 21 - 13333 \times 16\tfrac{1}{2} - 6666 (27\tfrac{1}{2} + 38\tfrac{1}{2}) + D \cdot 49\tfrac{1}{2}$$
$$(\text{rot. } r \cdot g).$$

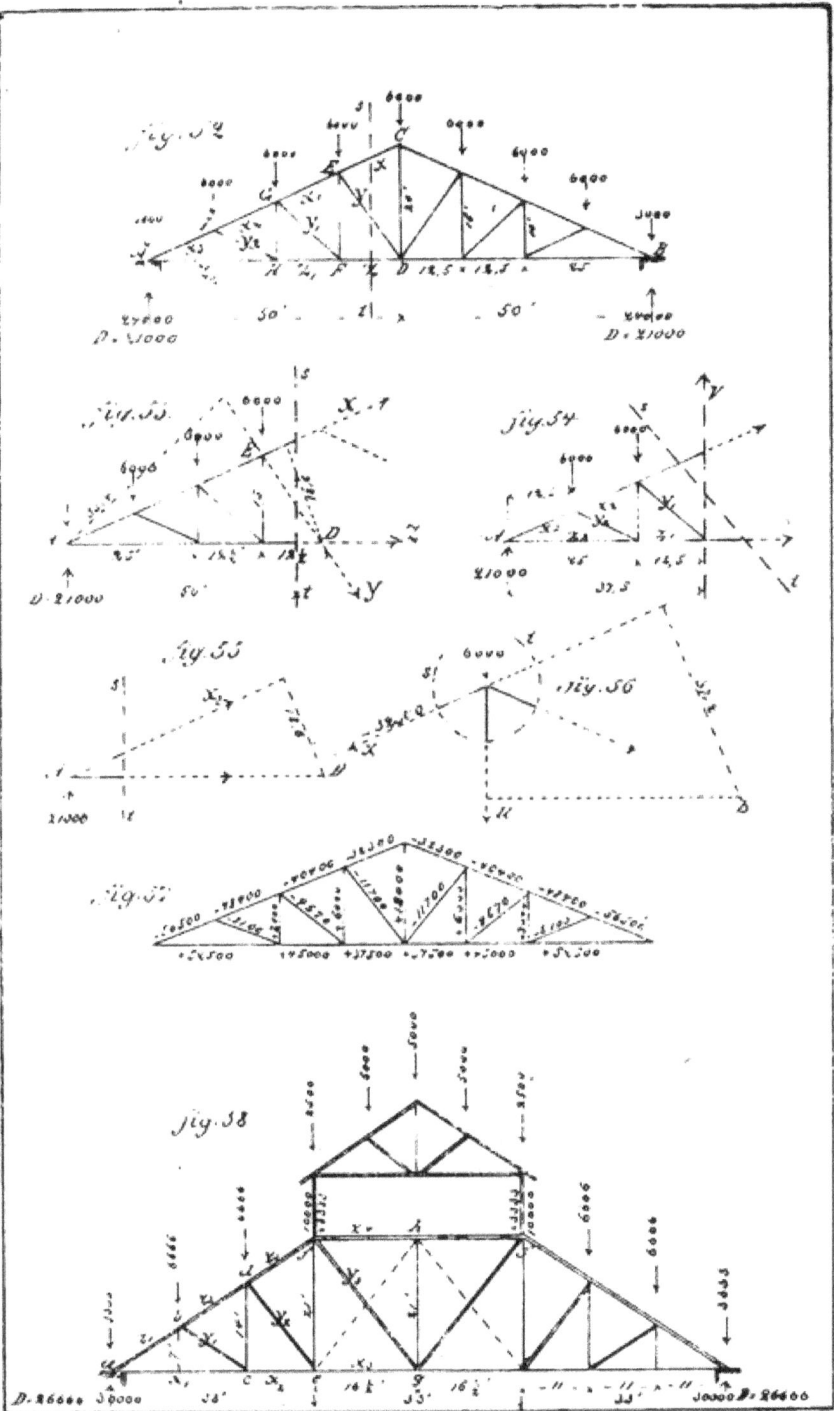

$$Z_4 = -\frac{659967}{21} = -31427,$$

and $\quad 0 = y_2 \cdot 39,5 + 13333 \times 33 + 6666 (11 + 22) \text{ (rot. } r \cdot a);$
$$Y_2 = -16708.$$

The tie-rod, gh, transmits the strain to the top flange, and is here sustained by the counter-brace, eh.

60.] From Fig. 60 is
$$0 = -x_2 \cdot 14 - 6666 \times 11 + D \cdot 22 \text{ (rot. } r \cdot d);$$
$$x_2 = \frac{513304}{14} = 36665;$$
$$0 = z_3 \cdot 17\tfrac{3}{4} - 6666 (11 + 22) + 26666 \times 33 \text{ (rot. } r \cdot c);$$
$$z_3 = -\frac{660000}{17,75} = -37180;$$
$$0 = y_2 \cdot 26,7 + 6666 \times 11 + 6666 \times 22 \text{ (rot. } r \cdot a);$$
$$y_2 = -8223.$$

61.] Fig. 61 gives the equations,
$$0 = -x_1 \cdot 7 + 26666 \times 11 \text{ (rot. } r \cdot b);$$
$$x_1 = +41902;$$
$$0 = z_2 \cdot 13 - 6666 \times 11 + 26666 \times 22 \text{ (rot. } r \cdot c);$$
$$z_2 = -39485;$$
$$0 = y_1 \cdot 13 + 6666 \times 11 \text{ (rot. } r \cdot a);$$
$$y_1 = -5640;$$

and for z_1 we find from the same figure,
$$0 = z_1 \cdot 13 + 26666 \times 22;$$
$$z_1 = -45125.$$

62.] For the strain in tie-rods we find from Fig. 62.
$$0 = -V_3 \cdot 33 + 6666 \cdot (11 + 22) \text{ (rot. } r \cdot a);$$
$$V_3 = +6666; \qquad \text{(Comp. Fig. 68.)}$$
$$0 = -V_2 \cdot 22 + 6666 \times 11 \text{ (rot. } r \cdot a);$$
$$V_2 = +3333;$$
$$0 = -V_1 \cdot 11 + 0 \text{ (rot. } r \cdot a);$$
$$V_1 = 0 \text{ (and is therefore not essential)}.$$

63.] The strain in V_4 at the centre rod, according to 8^b, can be defined thus:

$$V_4 = 2 \times \frac{21}{26,7} \times 16708 = 26200 \text{ lbs.}$$

The results are combined in Fig. 63.

64.] When in Fig. 64 the rafters are trussed—*i. e.*, stiffened by a king-post at b—there will be only four supporting points in the main construction, because the load in this case is transferred to the wall.

$$D = \frac{9999 \times 22}{99} + \frac{13333(33+66)}{99} + \frac{9999 \times 77}{99} = 23332 \text{ lbs.}$$

Plate 12.]
Fig. 65.] Further in Fig. 65,

$0 = x_2 \cdot 21 - 13333 \times 16\tfrac{1}{2} - 9999 \times 27\tfrac{1}{2} + 23332 \times 49\tfrac{1}{2}$
(rot. $r \cdot h$);
$\qquad x_2 = 31427;$

$0 = z_2 \cdot 21 - 13333 \times 16\tfrac{1}{2} - 9999 \times 27\tfrac{1}{2} + 23332 \times 49\tfrac{1}{2}$
(rot. $r \cdot g$);
$\qquad z_2 = -31427;$

$0 = y_2 \cdot 39,5 + 9999 \times 22 + 13333 \times 33$ (rot. $r \cdot a$);
$\qquad y_2 = -16708.$

66.] And from Fig. 66,

$0 = -x_1 \cdot 14 + D \cdot 22 = -x_1 \cdot 14 + 23332 \times 22$ (rot. $r \cdot d$);
$\qquad x_1 = +36665;$

$0 = z_1 \cdot 17\tfrac{3}{4} + 23332 \times 33 - 9999 \times 11$ (rot. $r \cdot e$);
$\qquad z_1 = -37180;$

$0 = y_1 \cdot 26,75 + 9999 \times 22$ (rot. $r \cdot a$);
$\qquad y_1 = -8223.$

67.] For Z we have from Fig. 67,

$0 = Z_1 \cdot 13 + 23332 \times 22$ (rot. $r \cdot c$);
$\qquad Z_1 = -39485.$

68.] See the results in Fig. 68 combined.

The strain $V_4 = 26200$ lbs. can be defined independently of the Method of Moments by the parallelogram of forces, as in Fig. 63, already shown,

$$V_4 = 2(-16708) \cdot \cos \alpha;$$

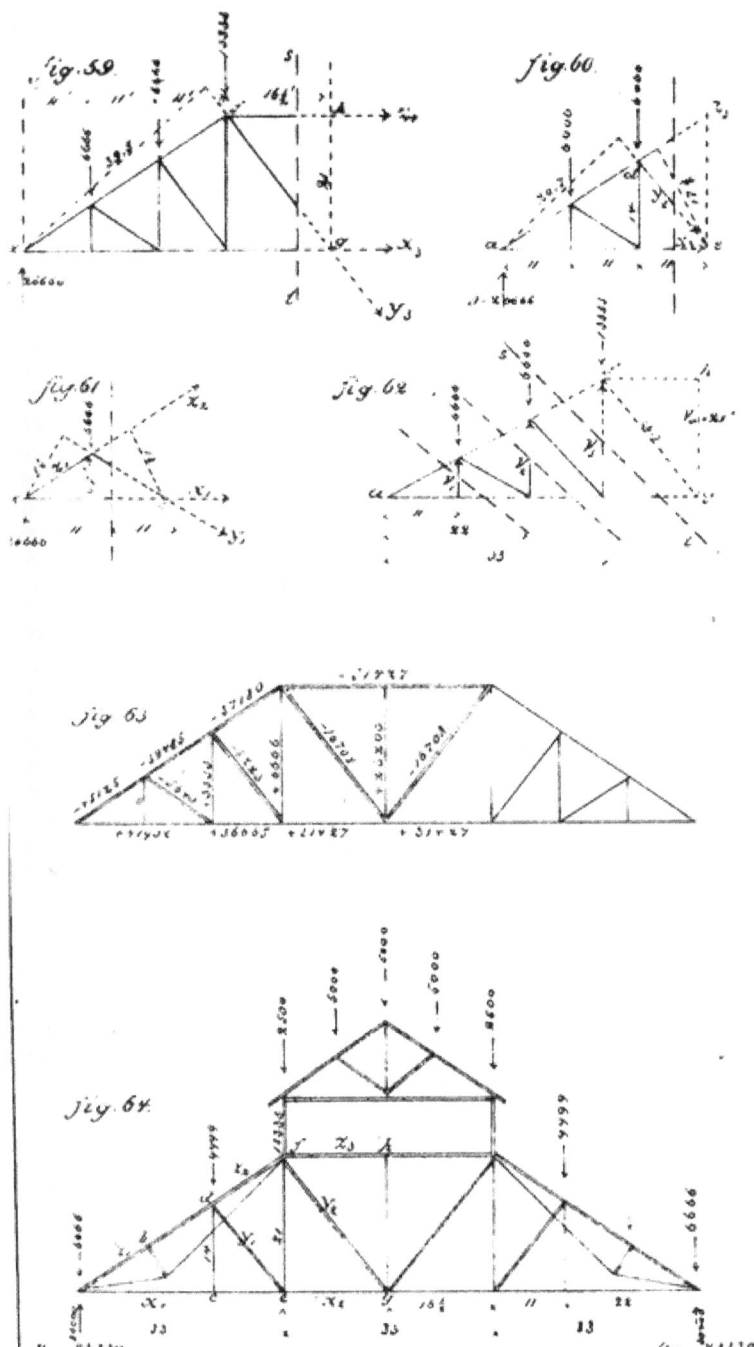

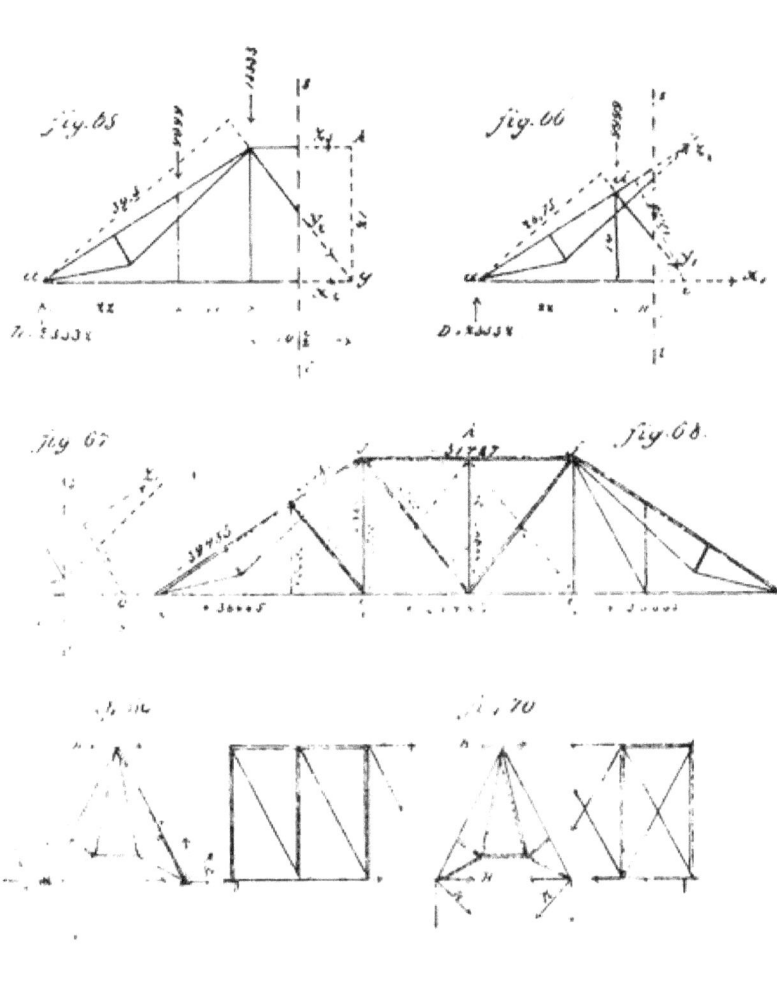

and when by means of counter-braces, e, h, the top chord is relieved from the strain, so that one-half to each side is transported to the tie-rods, e, f, then here the strain will increase to $13000 + 6666 = 19666$ lbs.

69.]
70.] In a combination of rafters (Figs. 69, 70), the pressure of the end rafters upon the wall results in an outward horizontal and vertical force.

Different from this is the action of the intermediate rafters, being similar to an oblique bridge-truss, sustained at the top chord.

The horizontal force at the heels of the intermediate rafters is opposed to the horizontal force of the end rafters.

[Plates 6, 7, 8, 9, 10, 11 and 12—embracing Figs. 31 to 70.]

C. SEMI-GIRDERS.

I. *SEMI-GIRDERS LOADED AT THE EXTREMITY.*

Plate 13,]
Fig. 71.] As the most simple presentation for a weight, W, the stress in struts and tie-rods is inscribed in Figs. 71 to 74, and the parallelogram of forces connected.

73.] To compute in Fig. 73 the stress in the lower flange, we have

$$\frac{de}{df} = \sec \alpha, \quad \text{and} \quad \frac{\frac{Z}{2}}{-W.\sec \alpha} = -\sin \alpha,$$

or
$$\frac{Z}{2} = -W.\sec \alpha.\sin \alpha;$$

$$Z = -W.\sec \alpha.\sin \alpha - W.\sec \alpha \sin \alpha,$$

or
$$Z = -2W.\sec \alpha.\sin \alpha;$$

and since
$$\sin \alpha = \frac{\tang \alpha}{\sec \alpha},$$

$$Z = -2W.\sec \alpha \frac{\tang \alpha}{\sec \alpha} = -2W.\tang \alpha$$

(much easier determined in Fig. 77 by the Method of Moments).

From Fig. 73 and the following we see that, for a load at the extremity, the diagonals are strained equally and alternately with tensile $(+)$ and compressive $(-)$ strains. (Comp. Fig. 23.)

But the strain in the flanges increases toward the support in each,
$$2W \cdot \tan \alpha,$$
where α is the angle of diagonals with a vertical line.

75.] For a better presentation of this, see Fig. 75, and for the calculation apply the Method of Moments.

76, 77.] When by a cut, st, a section of the structure is separated, we have for the flanges as equation of equilibrium (Figs. 76 and 77),

$0 = -x_1 \cdot cb + W \cdot ca$ (rot. r. b), | $0 = +z_1 \cdot h + W \cdot 2l$ (rot. r. d);

or for $cb = h$, and $ac = l$;

$x_1 = +W \cdot \dfrac{l}{h} = +W \cdot \tan \alpha$; | $z_1 = -2W \cdot \dfrac{l}{h} = -2W \cdot \tan \alpha$;

78, 79.] and by Figs. 78 and 79:

$0 = -x_2 \cdot h + W \cdot 3l$ (rot. r. e); | $0 = +z_2 \cdot h + W \cdot 4l$ (rot. r. f);

$x_2 = +3W \cdot \dfrac{l}{h} = 3W \cdot \tan \alpha$. | $z_2 = -4W \cdot \dfrac{l}{h} = -4W \cdot \tan \alpha$.

In the same manner is

$0 = -x_3 \cdot h + W \cdot 5l$ (rot. r. g); | $0 = +z_3 \cdot h + W \cdot 6l$ (rot. r. h);

$x_3 = 5W \cdot \dfrac{l}{h}$ | $z_3 = -6W \cdot \dfrac{l}{h}$.

CRANES.

Plate 14, Fig. A.] A wrought-iron crane (Fig. A), constructed of braces with link-joints, may be loaded at the extremity with 30000 lbs. $= P$; so is (for the dimensions noted in the skeleton) the horizontal strain, S, in a and b.

$$0 = -S \cdot 6 + P \cdot 12 \text{ (rot. r. } a);$$
$$S = 60000 \text{ lbs};$$

B.] and for the other members we have

$0 = -z_1 \cdot 0{,}75 + 0$ (rot. in the intersection of y_1 and x_1 or g, Fig. B);

$$z_1 = 0;$$
$$o = y_1 \cdot 1{,}1 - P \cdot 2{,}1 \text{ (rot. r. } i);$$
$$y_1 = +\frac{30000 \times 2{,}1}{1{,}1} = 57272;$$

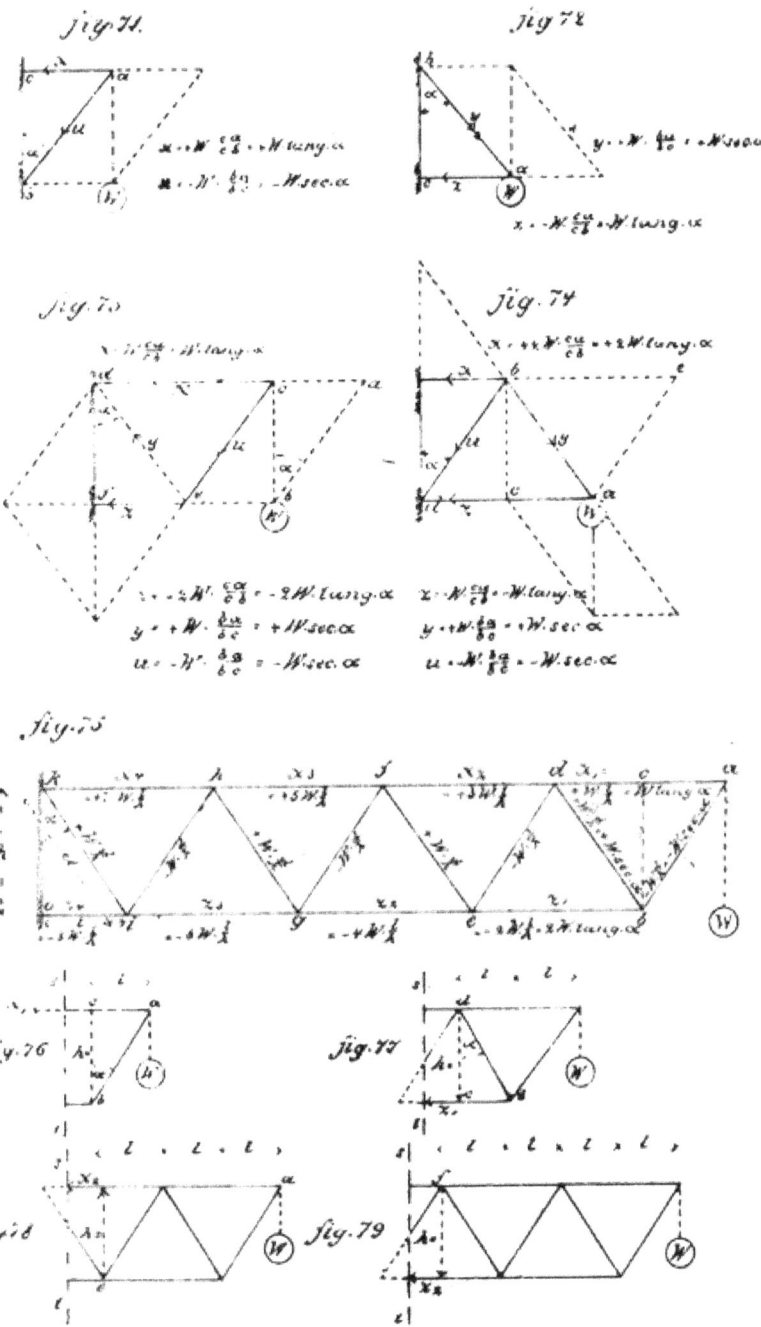

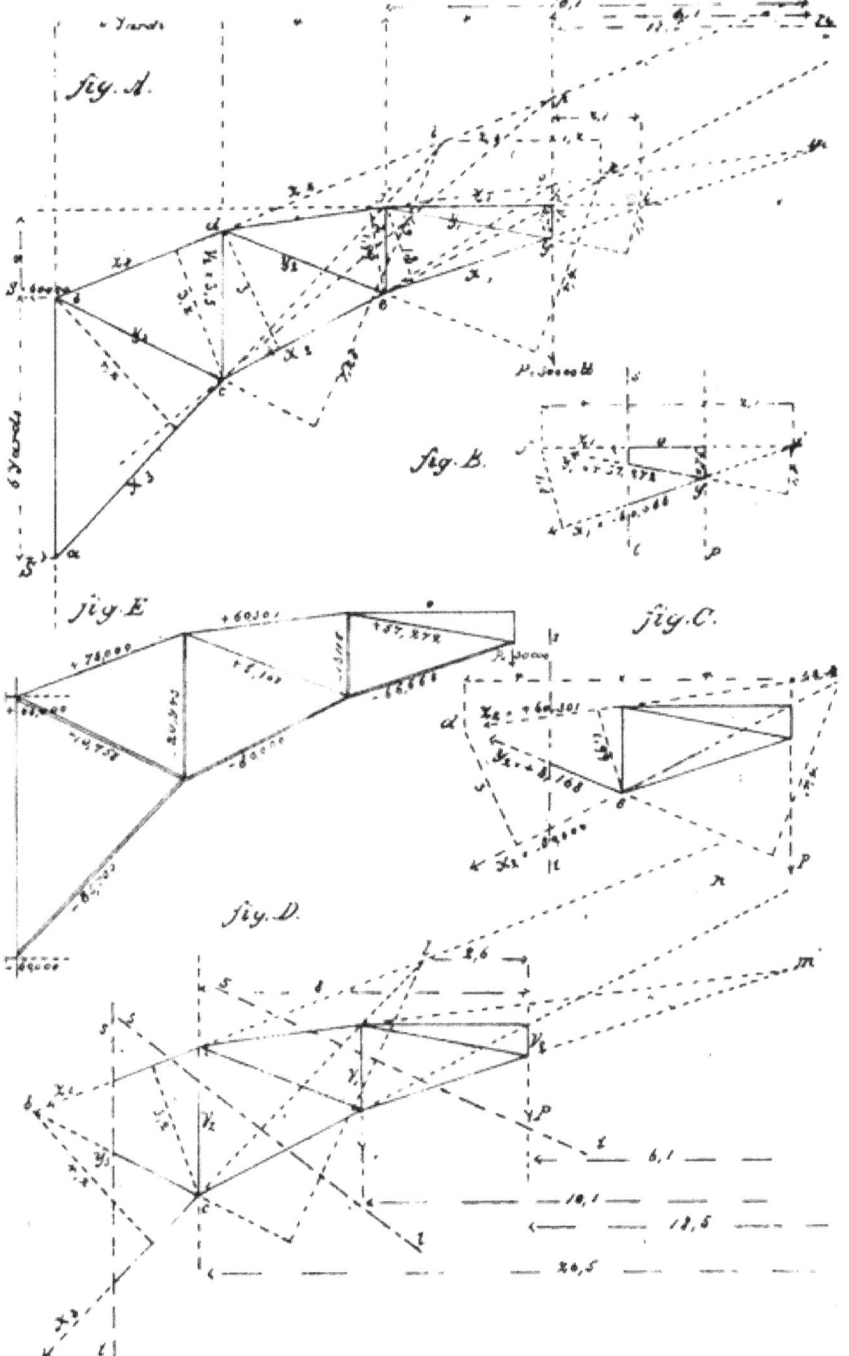

$$0 = x_1 \cdot 1,8 + P \cdot 4 \text{ (rot. } f\text{)};$$
$$x_1 = -\frac{30000 \cdot 4}{1,8} = -66666;$$

C.]
$$0 = -z_2 \cdot 1,99 + P \cdot 4 \text{ (rot. } r \cdot c, \text{ Fig. C)};$$
$$z_2 = \frac{120000}{1,99} = +60301;$$
$$0 = y_2 \cdot 4,4 - P \cdot 1,2 \text{ (rot. } r \cdot k\text{)};$$
$$y_2 = +8168;$$
$$0 = x_2 \cdot 3 + P \cdot 8 \text{ (rot. } r \cdot d\text{)};$$
$$x_2 = -80000;$$

D.]
$$0 = -z_3 \cdot 3,2 + P \cdot 8 \text{ (rot. } r \cdot c, \text{ Fig. D)};$$
$$z_3 = 75000;$$
$$0 = x_3 \cdot 4,2 + P \cdot 12 \text{ (rot. } r \cdot b\text{)};$$
$$x_3 = 85700.$$

For y_3 the intersection l of z_2 and z_3 is to the left of the suspended weight, and the symbol reversed.

$$0 = y_3 \cdot 7,25 + P \cdot 2,6 \text{ (rot. } r \cdot l\text{)};$$
$$y_3 = -10758,$$

which would be $= 0$ when the intersection is in the vertical line of the suspended weight, as the lines oe and pe in Fig. A indicate.

For the verticals, V, we have from Fig. D,

$$0 = -V_1 \cdot 10,1 - P \cdot 6,1 \text{ (rot. } r \cdot m\text{)};$$
$$V_1 = 18118;$$
$$0 = -V_2 \cdot 26,5 - P \cdot 18,5 \text{ (rot. } r \cdot n\text{)};$$
$$V_2 = -20943.$$

E.] The results combined in Fig. E.*

II. *SEMI-GIRDERS LOADED AT EACH APEX.*

In Fig. 25 is occasionally explained how to compute the stress in diagonals, as there is no intersection of joining flanges, x and z, and as in the case here considered the diagonals receive at each loaded

---

* For most purposes the above will be sufficient. In Glynn's rudimentary treatise on the Construction of Cranes we find valuable and complete drawings

Plate 15, Fig. 80.] apex an increment of strain, prior to the calculation may be given the general thesis that *the strain in two diagonals whose intersection is at an unloaded point is the same in numerical value, but of opposite character.* (Fig. 80.)

(See IV. General Remarks.)

The strain in diagonals, meeting at a loaded point, is in numerical value different.

The strain in flanges increases from apex to apex in geometrical progression.

81.] In Fig. 81 suppose the angle φ of diagonals with a horizontal line $= 45°$, so, also, angle $\propto$ $= 45°$; and by the table, sec $\propto = 1{,}414$.

When, again, in the axis $x = \infty$ a point of rotation, o, is supposed, we have per example for diagonal, y_3.

$$0 = + y_3 . x \sin \varphi + \left(W + W + \frac{W}{2} \right) . x \ (\text{rot.} \ r . o),$$

where $y_3 . \sin \varphi$ is the vertical component of y_3 or $= ab = c_1 b_1$ in the parallelogram of forces (Fig. 81), presenting by y, the resulting strain or diagonal. Divided by x, it follows:

$$0 = + y_3 \sin \varphi + \left(W + W + \frac{W}{2} \right);$$

and as $\varphi = 45°$, and $\sin 45° = 0{,}707$,

$$0 = + y_3 . 0{,}707 + \tfrac{5}{2} . W,$$

or $\quad y_3 = - 3{,}535 \ W.$

The same results from $- \tfrac{5}{2} W$. sec $\propto$, or $- \tfrac{5}{2} . W . 1{,}414$, which is also $= - 3{,}535 \ W$.

82.] In the same manner in Fig. 82 for y_2, the direction of which, when separated by a cut, st, is reversed to the weight. (See Figs. 22 and 23.)

$$0 = - y_2 . \sin 45 + \frac{W}{2} \ (\text{rot.} \ o);$$

$$y_2 = + \frac{\tfrac{1}{2} W}{0{,}707}, \quad \text{and} \ y_1 = - \frac{\tfrac{1}{2} W}{0{,}707};$$

and for the other diagonals,

$$0 = + y_3 . 0{,}707 + W + \frac{W}{2},$$

or $\quad y_3 = - \dfrac{\tfrac{3}{2} W}{0{,}707}, \quad \text{and} \ y_4 = + \dfrac{\tfrac{3}{2} W}{0{,}707};$

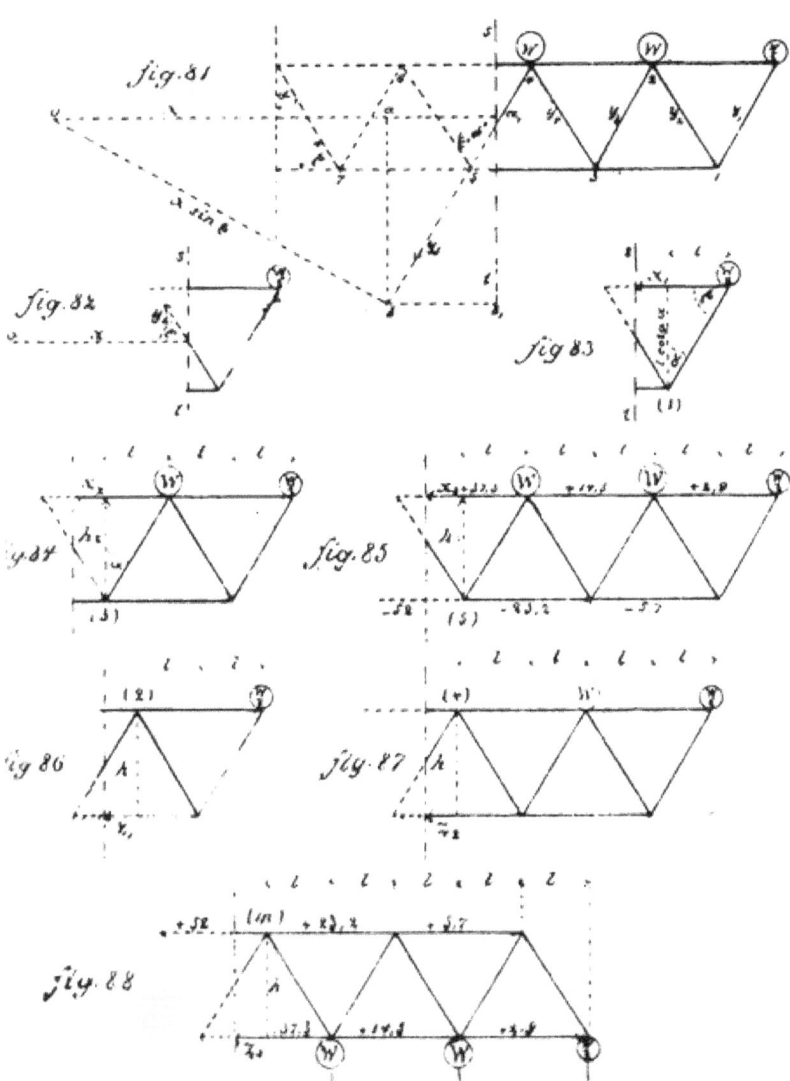

$$y_3 = -\frac{\frac{5}{2}W}{0{,}707}, \quad \text{and } y_4 = +\frac{\frac{5}{2}W}{0{,}707};$$

$$0 = +y_1 \cdot 0{,}707 + W + W + W + \frac{W}{2},$$

or $\quad y_1 = -\frac{\frac{7}{2}W}{0{,}707}, \quad \text{and } y_5 = +\frac{\frac{7}{2}W}{0{,}707}.$

83.] For the strain in flanges we have from Fig. 83,

$$0 = -x_1 \cdot l \cdot \cot \alpha + \frac{W}{2} \cdot l \; (\text{rot. } r \cdot 1);$$

$$x_1 = \frac{\frac{1}{2}W \cdot l}{l \cdot \cotg \alpha}, \quad \text{or as } \frac{1}{\cotg \alpha} = \tang \alpha,$$

$$x_1 = \tfrac{1}{2} W \cdot \tang \alpha;$$

and when $W = 10$ tons, $< \varphi = 60°$; therefore $< \alpha = 30°$,
and $\qquad \tang 30° = 0{,}577$ ("Example Stoney");

$$x_1 = 5 \times 0{,}577 = +2{,}9 \text{ tons};$$

or when, for an easier understanding, in Fig. 83,

$$l \cdot \cotg \alpha = h,$$

84.] we have for x_2 in Fig. 84,

$$0 = -x_2 \cdot h + W \cdot l + \frac{W}{2} \cdot 3\,l \; (\text{rot. } r \cdot 3);$$

$$x_2 = \frac{\frac{5}{2}W \cdot l}{h};$$

and as $\qquad \dfrac{l}{h} = \tang \alpha = 0{,}577,$

for our example,

$$x_2 = \tfrac{5}{2} \cdot W \cdot 0{,}577 = \tfrac{5}{2} \times 10 \times 0{,}577 = 14{,}5 \text{ tons}.$$

85.] So for x_3 in Fig. 85,

$$0 = -x_3 \cdot h + W \cdot l + W \cdot 3\,l + \frac{W}{2} \cdot 5\,l \; (\text{rot. } r \cdot 5);$$

$$x_3 = \tfrac{13}{2} \cdot W \cdot 0{,}577 = 37{,}3 \text{ tons},$$

and so further.

86.] For the strain, z, in the lower flanges,

$$0 = +z_1 \cdot h + \frac{W}{2} \cdot 2\,l \; (\text{rot. } r \cdot 2 \text{ in Fig. 86});$$

$$z_1 = -W \cdot \frac{l}{h} = -10 \times 0{,}577 = -5{,}7 \text{ tons}.$$

87.] Fig. 87 gives
$$0 = + z_1 \cdot h + W \cdot 2l + \frac{W}{2} \cdot 4l \text{ (rot. } r \cdot 4);$$
$$z_2 = -4W \cdot \frac{l}{h} = -4 \times 10 \times 0{,}577 = -23{,}2 \text{ tons.}$$

In the same way for z_3,
$$0 = + z_3 \cdot h + W \cdot 2l + W \cdot 4l + \frac{W}{2} \cdot 6 \cdot l;$$
$$z_3 = -9W \cdot \frac{l}{h} = -52 \text{ tons.}$$

88.] In case the load should be connected to the lower apexes (Fig. 88), the equation of equilibrium, per example for z_3, would be
$$0 = + z_3 \cdot h + W \cdot l + W \cdot 3l + \frac{W}{2} \cdot 5l \text{ (rot. } r \cdot m);$$
$$z_3 = -\tfrac{13}{2} \cdot W \cdot \frac{l}{h} = -37{,}3 \text{ tons;}$$

i. e., the strain is the same as in the flange of the reversed figure, but of opposite character. So also is the strain the same for the other flanges. (See Fig. 88.)

Remark.—In the given example the strains are determined without a certain length for h or l. This is easily explained by the relation which the angle φ or α bears to h and l, as by the extension of one, the other will increase in the same ratio.

[Plates 13, 14 and 15—embracing Figs. 71 to 88.]

D. GIRDERS WITH PARALLEL TOP AND BOTTOM FLANGES.

(Calculated for a Permanent Load.)

I. *STRAIN IN DIAGONALS AND VERTICALS.*

The calculation is very similar to the preceding. Provided, again, the load to be connected to the upper or lower apexes for the application of the Method of Moments, we now consider as a special force the reaction of the supports toward the system.

D_1 may represent one-half the weight of loaded truss or the pressure upon each support (prop), and diminished by the partition of load on this place directly sustained. (Comp. Fig. 21.) The reactive force of support wanted for our calculation will be signified by D.

To compute D, we refer to Fig. 4 in Sect. I., and define first for the following example its numerical value:

Plate 16,] Through bridge (over-grade bridge), between supports,
Fig. 89.]   48 feet;

8 panels, each 6 feet $= l$;

depth of truss $= 6$ feet $= h$, from centre to centre of top and bottom chords;

the weight of structure $= 3000$ lbs., and the load $= 15000$ lbs.;

gives a permanent load $= 18000$ lbs. per panel;

rolling load $= 0$. (See Sect. II.)

For the distribution of load, see Fig. 89.

Remark.—The strange impression which the arrangement of diagonals, unsymmetrical toward the centre, may first produce, will soon disappear after observation of the advantages for transformation upon succeeding systems.

Whole pressure of truss upon supports $= 8 \times 18000 = 144000$ lbs.;
$D_1 = 18000 (\frac{1}{8} + \frac{2}{8} + \frac{3}{8} + \frac{4}{8} + \frac{5}{8} + \frac{6}{8} + \frac{7}{8}) + 9000 \times \frac{8}{8} = 72000$ lbs.;
$D = 18000 (\frac{1}{8} + \frac{2}{8} + \frac{3}{8} + \frac{4}{8} + \frac{4}{8} + \frac{2}{8} + \frac{1}{8}) = 63000$ lbs.

The strain in the post, V_0, is 0, because $x_1 = 0$,
and $\qquad V_8 = -63000$.*

90.] Excepting the vertical component of y_1 (i. e., $y_1 \sin \varphi$), for a section (Fig. 90), only $D = 63000$ lbs. is a second vertical force.

Both turn to the left around o in the axis x_0; therefore their symbol, —. (Comp. Fig. 22.)

---

* For a deck-bridge (under-grade bridge)—i. e., when the upper apexes are loaded—would be
$\qquad V_0 = -9000,$ and $V_8 = -72000.$

The angle of diagonals with a horizontal line will be 45°.

www.ingramcontent.com/pod-product-compliance
Lightning Source LLC
Chambersburg PA
CBHW020226090426
42735CB00010B/1601